Of Sound Mind

Of Sound Mind

How Our Brain Constructs a Meaningful Sonic World

Nina Kraus

The MIT Press

Cambridge, Massachusetts | London, England

The MIT Press would like to thank the anonymous peer reviewers who provided comments on drafts of this book. The generous work of academic experts is essential for establishing the authority and quality of our publications. We acknowledge with gratitude the contributions of these otherwise uncredited readers.

This book was set in Stone Serif and Stone Sans by Westchester Publishing Services. Printed and bound in the United States of America.

Library of Congress Cataloging-in-Publication Data

Names: Kraus, Nina, 1952– author.
Title: Of sound mind : how our brain constructs a meaningful
   sonic world / Nina Kraus.
Description: Cambridge, Massachusetts : The MIT Press, [2021] |
   Includes bibliographical references and index.
Identifiers: LCCN 2020037100 | ISBN 9780262045865 (hardcover)
Subjects: LCSH: Hearing. | Sound—Physiological effect. | Brain.
Classification: LCC QP461 .K73 2021 | DDC 612.8/5—dc23
LC record available at https://lccn.loc.gov/2020037100

10 9 8 7 6 5 4 3 2

publication supported by a grant from
The Community Foundation for Greater New Haven
as part of the **Urban Haven Project**

To Mikey, Russell, Nick, and Marshall

Contents

Introduction

Of Sound Mind: A Partnership between Sound and the Brain

Sound Is Underrecognized; Hearing Is Underappreciated

It is a rare environment that is devoid of sound. Soundproof chambers exist that, on paper, are free of sound. But, if you have the opportunity to stand in one, you will quickly become aware of the slight rustle of clothing as you shift your weight from one foot to the other, the whisper of your own soft breathing, the soft thump of the beating of your heart, the creaking noises in your neck as you turn your head, the gentle scraping of your tongue as it brushes the back of your front teeth, your rumbling belly. Sound is all around us—inescapable and invisible.

Our sense of hearing is always "on." We cannot close our ears as we can our eyes. But, possibly more than any other sense, we are able to ignore sounds that are unimportant, to relegate them to the background of our consciousness. We have all experienced the sensation of becoming aware of a sound only after it suddenly goes away. Perhaps a refrigerator switches off. Or a nearby idling truck shuts off its engine. Or the downstairs neighbor turns off his television. The inescapability of sound along with our ability to tune it out makes our relationship with sound a complicated one. It is our primary means of communication and so is at the very core of our existence as interconnecting humans. Yet hearing is often taken for granted. Most of us, facing the dilemma, would give up our hearing before we would give up our vision because we can imagine navigating our daily lives in silence, but not in darkness. Sound is underrecognized. Hearing is underappreciated.

My interest in sound began early. I grew up with music—my mother was a pianist. My favorite place to play, as a child, was under the piano. I would bring my toys there and play my games against the backdrop of Bach, Chopin, and Scriabin. I also grew up in a house where more than one language was spoken as we traveled back and forth between New York and la mamma's native Trieste, Italy. I had friends and family in both countries and navigated both languages pretty well. These early experiences with language and music left a deep imprint on me and are why, years later, as a neuroscientist and college professor, my favorite course to teach is The Biological Foundations of Speech and Music. That course and this book are about sound—its richness, its meaning, its power—and about the brain that makes sense of it all, making us who we are.

The path between la mamma's piano and studying the exquisitely precise auditory brain as it processes the sounds of our lives was not a straight line. In college, my interest in words and languages led me first to comparative literature. That was my major until I took a course in biology. Around the same time, I found a book by Eric Lenneberg, *Biological Foundations of Language* (sound familiar?).[1] In it, Lenneberg wrote about biological and evolutionary principles that make language possible. It married the study of language with the study of biology in a way that was new at the time. That got my attention. I realized this area of study was possible, and I knew it was the one I wanted to pursue. But I did not want to limit myself to language. I was interested in the broader topic of sound itself. Sounds are all around us on the *outside*, but what goes on *inside* the brain when we hear a word or a chord or a meow or a screech? How do sounds change us? How do our experiences with sound change how we hear it? I had homed in on the biology of sound processing as a field of study.

When I got to graduate school, I realized I could get paid to *learn*. My monthly stipend was $200 and my rent was $50. I was set! Now I just had to figure out what route my pursuit of the biology of sound processing would take. I soon found myself in a lab studying two-tone suppression in the auditory nerve of the chinchilla—the influence one sound has on another when both sounds occur simultaneously.[2] As I was enthusiastically explaining all of this to la mamma, she looked at me and asked, "Nina, what are you doing?" At that moment, I realized I could not explain why two-tone suppression in the chinchilla should matter to her. Why would I want to research it? Nina, what are you *doing*?

It became clear that if I couldn't explain to my mother how I was spending my time, I didn't want to be spending my time that way. I realized the science I do needs to be explicitly grounded in the lived world. I was still immensely interested in sound and the brain, so I moved on to my next lab, where I worked with rabbits and the auditory cortex. There I discovered that with training—learning to assign meaning to a sound—the individual neurons in the auditory brain changed their behavior.[3] When a sound has little meaning, the brain will respond to it one way. But when that same sound has acquired relevance—food is on the way, for example—the brain responds differently. A sound-brain partnership is formed, connected to the living world. The *meaning* of the signals outside the brain matters to the signals inside the brain. This was news at the time and, more importantly, it was something I could explain to la mamma. She could see the significance of it—*anyone* could. I intended to find out how and why the brain changed its response to a sound that had meaning.

Sound Connects Us to the World

The ability to perceive sound is evolutionarily ancient. All vertebrates have a mechanism of hearing. In contrast, many vertebrates are blind, including some mole, amphibian, and fish species, and a host of cave dwellers. Sound perception evolved for self-preservation, a warning system against predators or other environmental dangers. The stressful feeling the clanging roar of traffic causes may be a twenty-first-century shadow of our distant ancestors reacting to a noise signaling a coming avalanche or stampede.

Helen Keller commented that "blindness disconnects us from things; deafness disconnects us from people." Sound represents things we cannot see and cannot describe. Think of how your mother asks, "What's the matter?" the instant she picks up the phone and hears your not-quite-right-sounding voice. Sound is unseen but is palpable and dense with meaning.

Why then does vision come out on top in a "favorite sense" poll?[4]* Why was the National Institutes of Health's institute for vision founded twenty years before one devoted to hearing? I think one reason is that we have forgotten *how* to listen. The constant racket around us has made us numb to sound and incapable of hearing sound details. We choose then to ignore sound and turn instead to vision. Another reason is that, like gravity and other powerful forces in our lives, sound is invisible. When was the last time you really paid attention to gravity? Out of sight, out of mind. Finally, sound is fleeting. If we see a tractor lumbering through a cornfield, it remains large, yellow, and metallic even as it passes from one side of our visual field to another. It has a permanence. It waits for us to soak in its tractorness and rewards our extended, leisurely viewing with a palate of sight-related descriptors.

---

*Some 2,000 US adults responded to an online poll asking them to rank the disease or ailment that was "the worst that could happen to them." Blindness was ranked the worst, trouncing deafness and a number of other fairly dire things including Alzheimer's disease, cancer, and loss of a limb.

But a sound can be over in an instant or evolve over time into a different sound in a flash. And once it's gone, it's gone.

Consider the smallest unit of speech from an acoustic standpoint. The word "brink" has only one syllable, but it has five discrete phonemes or unique sounds. Change any one of them and the meaning is changed ("drink") or lost ("brint"). In running speech, we hear as many as twenty-five to thirty phonemes *every second*, and if we do not process them properly, the message may be lost. But, in most circumstances, this swirl of sound poses little challenge to our speedy auditory systems. Think about having to process a *visual* object that changes twenty-five to thirty times in a second. There's a ball! Now it's a giraffe! Now it's a cloud!

How do we manage to identify speech that is moving much too fast to leisurely study? We harness the unmatched speed and computational power of the auditory brain. Think about how long a second is. Now think about a tenth of a second. Now a hundredth of a second. At that point, it's pretty hard to even comprehend how fast that is. Now add another zero. Auditory neurons make calculations at one thousandth of a second. Light is faster than sound, but in the brain, hearing is faster than seeing, touch, and any other sense.

Our Hearing Brain Includes Sensing, Moving, Thinking, and Feeling

We do not just *hear* sounds; we deeply *engage* with them as we make sense of sound. *Our hearing brain is vast.* Hearing involves sensing, moving, thinking, and feeling. Until recently, we didn't see it this way.

The beautiful, specialized auditory structures that connect the ear with the brain may at first bring to mind workers on an assembly line. A product (sound) enters the ear and is moved from station to station, picking up parts along the way. This hierarchical, one-way portrayal is the classical view of sound processing. It still persists but is a gross simplification and misses the big picture. The auditory pathway is not a one-way street in the middle of a desert; it's part of an all-way super-highway in a busy urban center, complete with on- and off-ramps, traffic circles, and spaghetti bowl interchanges, routing traffic to and from many brain neighborhoods. When it is all operating at peak efficiency, it is a wonder of infrastructure and traffic flows smoothly and speedily. But like an urban highway, there can be a backup caused by an incident a mile away in a part of town that has no obvious bearing on the bumper-to-bumper traffic I am experiencing right now.

Yes, there are hierarchies, compartments, and specialties within the auditory pathway, but they are important to the extent they interconnect and connect with forces outside themselves. Human achievements like speech and music did not come about from the auditory-processing centers dutifully moving information about the auditory soundscape one-way from the ear to the brain. Rather, these achievements are a result of a deep network of interconnectivity between our sensory system, the motor networks, the system that drives motivation and feelings of reward, and cognitive centers that govern how we think. Indeed, hearing involves sensing, moving, thinking, and feeling (figure I.1).

Auditory-*motor* connections enable us to move our mouths, tongues, and lips to speak and sing, and work closely with various parts of our bodies when we play musical instruments. When we listen to speech, we unconsciously move our tongues and other articulatory muscles in synchrony with whom we are listening to.

Hearing and thinking are linked. We may have certain instinctual vocalizations—the sound that emerges when I hit my finger with a hammer comes to mind. But a great deal of *cognitive*, intellectual capacity is required to speak even the simplest sentences or play the most basic music. And it cuts both ways. The risk of dementia is significantly higher in people with hearing loss. It is not just that hearing loss makes it harder for Uncle Joe to follow conversations and therefore seem not as with-it. Hearing loss impairs how we think.[5]

The sound of speech and music has privileged access to the brain's *reward*, or emotional, network. Speech and music might not

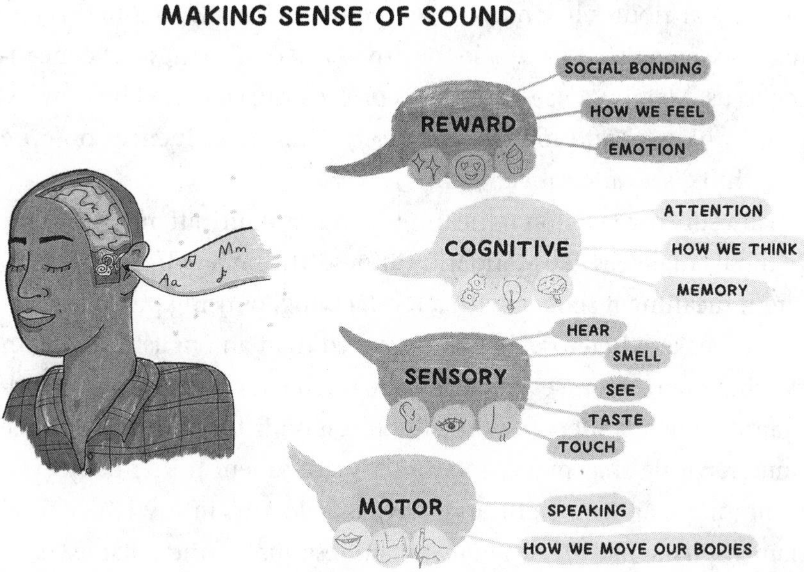

MAKING SENSE OF SOUND

REWARD
SOCIAL BONDING
HOW WE FEEL
EMOTION

COGNITIVE
ATTENTION
HOW WE THINK
MEMORY

SENSORY
HEAR
SMELL
SEE
TASTE
TOUCH

MOTOR
SPEAKING
HOW WE MOVE OUR BODIES

Figure I.1
Making sense of sound engages how we think, feel, sense, and move.

have evolved if not for the deep emotional feelings of connection with other humans that arise during these communal activities. Indeed, sound contributes to our sense of belonging to the world, to our own personal sense of home.

That hearing does not take place in an isolated, one-way path is now largely accepted, but the shift in thinking this statement represents is relatively new—within the span of my career. The interconnectedness of the auditory system with the rest of the brain has a dramatic effect on how we process sound. It is the heart of our experience with sound, with people, and our individuality.

The Hearing Brain Is Shaped by Experience

My husband and I frequently disagree about the thermostat setting because we experience the same temperature differently. Sensory systems are not scientific instruments that objectively measure physical attributes like mass or temperature. Instead, our brains format the signals that comprise the physical world so they have meaning to us. Making sense of sound is profoundly governed by how we feel, think, see, and move. Conversely, hearing influences how we feel, think, see, and move.

I am sure my reaction to hearing "Nina" is quite different to yours. In tonal languages like Mandarin Chinese, the same syllable has a different meaning if spoken with a level, falling, or rising pitch. A Mandarin speaker, therefore, is more invested than an English speaker in tapping brain resources to code these pitch cues.[6] Over time, sound-brain teamwork alters how the brain responds to sound. This is the same rewiring that makes mommy's voice salient to her baby even if mommy is not in sight and why, anecdotally, in my lab, a child named *Day*na had an extra-large brain response to the syllable "day" compared to the "doo," "doh," "dah," and "dee" syllables she also listened to in one of our experiments (figure I.2).

Figure I.2
Sound processing in the brain is affected by the languages we speak, the music we make, and our brain health.

No Boundaries

When I was five years old, the neighborhood kids said, "You have to be six to play with us." This interaction and others like it, along with spanning two cultures—not feeling fully Italian nor fully American—has long made me wonder where I belong. Where do I belong as a scientist? I have always felt most comfortable at the intersection of disciplines, rather than squarely at the center of one, and so I have constructed my lab, Brainvolts, in that image.

If you look at Brainvolts' website, you will see music, concussion, aging, reading, and bilingualism among the areas we investigate. You might ask, "Just what *are* they doing at Brainvolts?" The simple, unifying theme is the sound-brain partnership. Sound cuts across many aspects of our lives and shapes our brain accordingly.

My husband refers to Brainvolts as my "hot dog stand." It is my job to do what it takes to create the necessary infrastructure to sell hot dogs. A scientist needs specialized equipment and, most of all, she needs the right people. It can be agony because my interests rarely thread the specialty purview of most funding sources. I often feel like I'm five again, hearing, "We only fund six-year-olds." That is the agony of operating across borders although, thankfully, I've managed to keep the hot dog stand turning out hot dogs. On the ecstasy side, science has brought me into the orbit of exceptional people outside research and academia. The science is first and foremost grounded in the people at Brainvolts who bring their unique perspectives to our common purpose. Our science depends on our collaborators in education and music and biology and athletics and medicine and industry—people who operate in the world outside the lab, the world I want our science to live in. As neuroscientist Norm Weinberger put it, "Nature doesn't respect disciplines."

Brainvolts, much like the brain, is an integrated and reverberating system-wide network, connected by unique and specialized individual parts—er, team members. Since its inception some thirty-odd years ago, I have been uncommonly fortunate to work with outstanding individuals who bring their own interests, points of view, and skills into the lab, each with an abiding interest in the interface of sound and the brain. We will explore these networks—both in the brain and at Brainvolts—in the coming pages.

The Sound Mind

As this book began to take form, I shot early drafts to friends and family for feedback. I wanted to know if my writing was understandable and the topic of interest to a cross-section of readers. My immediate family, conveniently composed of a chef, a lawyer, a carpenter, a musician, and an artist, bore the lion's share of this. Fairly early on, my lawyer son-in-law asked whether this book is about *sound* or about the *brain*. This made me want to specify the answer is both. It is about sound, what our brains do with it, and also what this does to us—the *sound mind*.

Said another way, I think of the *sound mind* as a force behind a continuum from the past to the present and into the future. The sounds we have engaged with over our whole lifetimes have shaped what our brain is today. Our brain today, in turn, can make decisions about how we shape our sonic world going forward, not just our personal futures but those of our children and of society as a whole. Thought about this way, the sound mind drives a feedback loop that, importantly, we have some control over. We have the power to make choices about sound for better or worse. Will we make the right decisions to make the feedback loop a virtuous circle? Or make poor decisions launching a vicious circle?

As a biologist, I want to know how sound develops our sonic personality and enables us to engage with our world. I aim to understand sound processing in the brain—the sound mind—with the precision I've experienced recording directly from individual neurons.

This book will examine signals outside the head (sound waves) and signals inside the head (brain waves). We will look at ways we can enrich sound processing, and the mechanisms through which processing can be adversely impacted. We will consider the power of music for healing as well as the destructive power of noise on the nervous system. Along the way, we will cover what happens to the sound mind when we speak another language, have a language disorder, experience rhythm, birdsong, or a concussion.

Sound is an invisible ally and enemy of brain health. Our engagement with sound leaves a fundamental imprint on who we are. The sounds of our lives shape our brains, for better and for worse. And our sound minds, in turn, impact our sonic world, again for better and for worse. Will we be expert listeners or poor listeners? As a consequence of what we value in sound, how will we build the sonic world we live in? A holistic understanding of the biological consequences of our lives in sound positions us to make better choices for ourselves, for our children, and for society.

I'd like to think la mamma would have enjoyed reading this book.

I   How Sound Works

1

Signals Outside the Head

This opening chapter is about the signals that are found outside our heads—*sound*. Sound is just air molecules moving back and forth. Remarkably, from this simple mechanism comes an infinite variety of sounds, from Bach to bacon sizzling, from Rocky Raccoon to the raccoon out back foraging in the garbage can. Sounds can be loud or soft, high or low, consonant or dissonant, fast or slow, rough, reedy, chaotic, polyphonic, whooshing, and staticky. I invite you to savor the beauty of the properties of sound—the ingredients we will return to time and again as we explore the sound mind.

Sound is movement. When a guitar string is plucked, it moves the air nearby. Figure 1.1 shows a guitar string in different states of pluckedness. On the left is a guitar string at rest, with a dozen little air molecules hanging out to its right. When the guitar string is at rest, the local atmospheric pressure is around 14.7 pounds per square inch—the air pressure at sea level. When the guitar string is plucked, it briefly moves to the right and our air molecules are squished closer together—that is, they are compressed to a higher pressure.* Then,

---

*This change in pressure is infinitesimal. If I got my math and my unit conversions correct, an actual pluck of a typical guitar string will raise the local atmospheric pressure from 14.7 psi to something like 14.700003 psi.

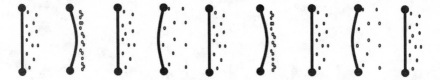

Figure 1.1
A plucked string moves the air molecules around it.

after a very short time (hundredths or thousandths of a second, depending on the pitch of the note), the guitar string springs back in the direction of its resting position and continues moving past the initial resting state until it is displaced a bit to the left. Then the air molecules on the right are spread out again, reducing pressure. But they don't fall right back into the same spacing as before the string was plucked. They overshoot a bit, so they are now spread out more—they are at a lower pressure—than they were before the string moved in the first place. Then they rebound together again, and spread out again, and so on, a little less each time, until eventually the movement stops, the vibration dampens to nothing, and the sound dies out. The movement was the sound, and when the movement stops, the sound is over.

Sound Ingredients

Most sounds can be described by a handful of *sound ingredients* (figure 1.2), much as a seen object can be classified by shape, color, texture, and size. Because sound is invisible, the ingredients are not as obvious, but they are crucial to how we make sense of sound. Thinking about sound in terms of its constituent ingredients— recognizing the wealth of what is going on in those moving air molecules—makes its processing in the brain even more amazing in my view. To keep tabs on those marvelous ingredients, I find a helpful organizing principle, then, is to think of sound in terms of *pitch*, *timing*, and *timbre* (pronounced *tamber*).

Figure 1.2
The endless variety of sound arises from air movement and can be described by a handful of ingredients.

Pitch

Pitch is the perception of "high" vs. "low." We describe a flute's sound as high-pitched and a tuba's as low-pitched. What we hear when we use those labels arises from the physical property of *frequency*. We hear a high-pitched sound when the fluctuations between high and low air pressure come very fast, or at a high frequency. A low-pitched sound has more leisurely changes in air pressure—a low frequency (figure 1.3). Pitch is a perception; frequency is a physical, measurable property. We should use care in making this distinction between pitch and frequency because they are not always a perfect match.

Frequency—not as a scientific measure of sound but as an English word—means a count of some event with respect to a fixed time period. You might get two paychecks per month. Tampa, Florida, has on average seventy-eight thunderstorms per year. I get twenty-two

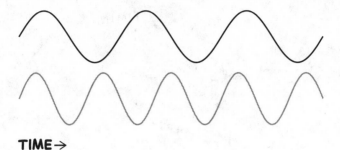

TIME→

Figure 1.3
The gray waveform has more cycles (is at a higher frequency) than the black, and so would sound higher in pitch.

pieces of junk mail per week. These are all frequencies. The number of air-pressure vibrations per second distinguishes the pitch of a flute from the pitch of a tuba. The term for the count of something per a time unit of one second is hertz, abbreviated Hz. The range of air pressure fluctuation frequencies that a human ear can detect is between 20 Hz and 20,000 Hz. A high-pitched flute can play notes with frequencies in the range of about 250 to 2,500 Hz; a low-pitched tuba 30 to 380 Hz. Surprisingly, there is a bit of overlap in their ranges! I think I will get to work on writing a flute and tuba concerto where the tuba has the higher part.

But there isn't always a perfect match between the frequency of a sound and the pitch we hear. If a sound is perceived as having a pitch—if is "hummable"—the frequency at which we would hum is called the *fundamental frequency*. In figure 1.4, both waveforms have an identical number—about thirty-five—of peaks and valleys, so they are nominally the same frequency. However, each is turned on and off—modulated—at different rates. The pitch we hear matches the rate of the modulation, not the frequency of the wave that is being modulated.

An example of this is the human voice. The pitch (fundamental frequency) of the human speaking voice ranges from about 50 to 300 Hz. In speech, the fundamental frequency corresponds

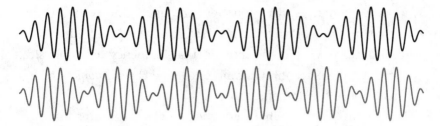

TIME →

Figure 1.4
The black and gray waves have the same frequency. But the modulation rates are different—that is, the sound is being turned on and off at a faster rate in the gray waveform and so would sound higher in pitch than the black. The faster modulation rate in females—provided by faster vibration of the vocal cords—produces a higher voice pitch when the same words are spoken.

to the speed of the openings and closings of the vocal folds set in motion by our breath. The speed of vocal fold movement is slowest for men, resulting in a deep voice, and highest in children, resulting in a high voice. Interestingly, voice pitch differs not only between individuals and sexes but in some other surprising ways. Fundamental frequency differences have been observed, on average, between speakers of different languages[1] and between demographic groups within the same language.[2] And we may have observed, perhaps even in ourselves, a bilingual person who speaks at a generally higher pitch in one of their languages than the other.[3]

Timbre

In music, timbre is the primary means we use to distinguish two instruments playing the exact same note. In speech, it is our primary cue for distinguishing one speech sound (consonants and vowels) from another. A man and a woman say the same thing: the fundamental frequency (voice pitch) helps us determine who is who. A woman says two different things: timbre is the distinction that helps us distinguish her "so" from her "sue." Just as the perception of pitch has the fundamental frequency as its physical counterpart, the perception of timbre is defined by *harmonics*, the frequencies above the fundamental.

It is useful to know what frequencies a given sound is made up of. This is known as a sound's spectrum. A tuning fork's spectrum is made up of one and only one frequency, so it has a single, thin, vertical line as seen in the top panel of figure 1.5. It has no harmonics, just a fundamental frequency. A natural sound such as a middle C played by a trombone or a clarinet will likewise have a peak in the spectrum at middle C's fundamental frequency of 262 Hz plus additional peaks at multiples (524, 786 . . .) of the fundamental. These are *harmonics*. It is apparent in the middle and lower panels of figure 1.5 that not all harmonics have the same amount of energy. The patterns of relative energy levels are signatures of the trombone and clarinet and why we can hear the difference between them. The unique harmonic signatures are determined by the shape and construction of the instrument producing the sound. Analogously, the shape and position of our tongue, mouth, and nose produce the harmonic patterns that distinguish different speech sounds.

Depending on the position of our lips and tongue and the amount of air that gets routed through our nose and mouth, we alter the spectrum (which harmonics get reinforced) as seen in figure 1.6. While the spectra of the two vowels have peaks every 100 Hz (due to a fundamental frequency of 100 Hz in this example), the relative sizes of the peaks, outlined by the gray lines, are very different. This is the speech analog of the trombone/clarinet distinction. For "ee," the two bumps in the gray line come at about 300 and 2,300 Hz; for "oo," they appear at about 400 and 1,000 Hz. Speech contains bumps in the spectrum—areas of maximal energy concentration (called formants). Interestingly, these bands of acoustic energy are reasonably similar among speakers. A talker with a high voice pitch

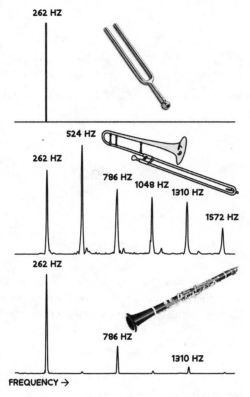

Figure 1.5
A spectrum of a tuning fork is a single vertical line at only one frequency, here 262 Hz, or middle C. An instrument playing a middle C will have a peak at 262 Hz plus several harmonics at multiples of 262 Hz. A middle C played by a trombone or a clarinet features different patterns of harmonics due to resonance characteristics of the instruments. The spectra help us see why the same middle C sounds different when played by different instruments. (Frequency on the x-axis; energy on the y.)

will have peaks somewhere in the neighborhood of 400 and 1,000 Hz in her "oo" just like a low-pitch speaker.

So *timbre* is the perception that arises from the *harmonic* content in a sound. Harmonics—where they appear and how big they are relative to one another—are the physical attributes of sound that enable us to tell, by their timbral quality, the difference between two instruments or two speech sounds. In speech, groups of harmonics

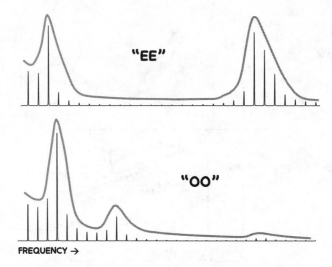

Figure 1.6
Top: spectrum of "ee" as in beet. Bottom: spectrum of "oo" as in boot. Both have the same fundamental frequency, but where the energy of the harmonics is concentrated differs. (Frequency on the x-axis; energy on the y.)

stand out within the spectrum of a particular word or syllable. Figure 1.7 illustrates the full frequency range (fundamental and harmonics) of a few instruments and voices.

Timing

Until now, we have been discussing tuning forks, single musical notes, and vowels—all examples of sounds that are stable over some period of time. But there is a class of sound where timing is a defining characteristic of the signal itself—not in terms of when the sound starts and stops, like syllables or musical notes, but when and how the sound itself evolves over time. Among these is the speech consonant. In certain consonants, timing takes center stage.

Say the word "bill" out loud. Now say the word "gill." Can you describe what mechanically differed between the two inside your mouth? Easy peasy. In the first case, your lips came together and your

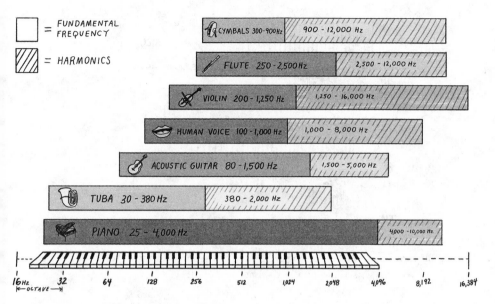

Figure 1.7
The full frequency ranges of musical instruments and voices. The fundamental frequency range is shown on the left. Harmonics are shown to the right.

tongue was in a somewhat neutral position. In the second, your lips were slightly open and you pressed the back of your tongue to the roof of your mouth. Now say "bill" and "pill." This one is trickier. What exactly is different? The salient mechanical difference between a "b" and a "p" may not be immediately obvious. Your tongue and lips are pretty much in exactly the same position for both. The primary difference is in the timing—*when* you start voicing the vowel—that is, when your vocal folds begin buzzing out the "i." In "bill" you start voicing the vowel right away. But in "pill" you wait a very short amount of time after your lips part before you begin to voice the vowel. In the top wave of figure 1.8, you can see the sound wave of the word "bill." In the bottom wave, I have inserted 1/20 of a second of silence. Every single wiggle is identical between the two except for the added silence. That little gap before starting to voice the "i" sound is enough to make the second one very clearly sound like "pill." A timing cue of a fraction of a second makes a big difference in

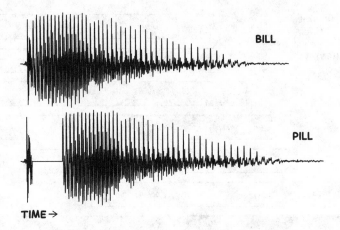

Figure 1.8
"Bill" turns into "pill" by adding 1/20 second of silence just before the vocalization of the vowel begins. (Time on the x-axis; energy on the y.)

language. This is one of the many reasons you and I need a super-fast auditory brain to process such tiny changes in sound.

Looking at Frequency Changes Over Time

Differences in *timing* such as "bill" vs. "pill" are fairly easy to see in time plots like figure 1.8. Differences in *frequency* such as "ee" vs. "oo" are fairly easy to see in spectrum plots like figure 1.6. However, neither plot does justice to the acoustic distinction between a "b" and a "g." This involves a change in *frequency that unfolds over time*. To adequately depict a "b" vs. "g" difference, we need the third and final plot, the *spectrogram*.

The top panel in figure 1.9 is a simple example, showing a tone that, over time, goes from low frequency to high, then back down to low—like a stereotypical wolf whistle. Imagine a siren or sweeping your finger across the notes on a piano.

In consonants like "ba" and "ga," bands of acoustic energy sweeping across frequencies drive the distinction (bottom panel). The upper band is the same in "ba" and "ga," a harmonic band moving in time from lower to higher frequency until it flattens out at the "a."

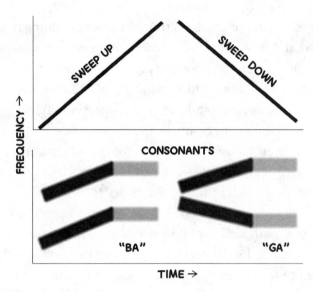

Figure 1.9
Spectrograms (depicting changes in frequency over time). Top: an upward, then downward sweep in frequency. Bottom: "ba" and "ga." The frequencies of both bands of acoustic energy change over time until they stabilize at the vowel "a."

But the bottom band differs for the two syllables. For "ba" it moves from low to high frequencies before leveling off. For "ga" it starts higher and moves downward in frequency. The term *FM sweep*—an important ingredient of sound—refers to this type of change in frequency over time.

So, in both of our consonant-pair examples, "b" vs. "p" and "b" vs. "g," timing is a crucial component to identity. In "ba/pa," timing is both necessary and sufficient to set up the contrast. In "ba/ga," the interplay of *both* time and frequency makes the distinction. While we can capture and isolate these sound distinctions by slowing the sound way down and measuring them, in practice, they happen far too quickly for us to consciously perceive what is driving the distinction. Remarkably quickly. Think about it: did you know the difference between ba and ga in terms of sound ingredients before I told you? Did you realize that a couple of blink-fast FM sweeps can turn a muddy dog into a muggy bog? I certainly cannot tell by listening that a certain band of energy is rising in a ba and falling in a ga. Yet this speed and subtlety makes consonants perceptually vulnerable, necessitating the use of phonetic alphabets (alpha, bravo, charlie, delta . . .). The subtlety and complexity of these distinctions and the difficulty some people have processing them have intriguing consequences for language and even reading, as we shall see.

We have been focusing on speech for our discussion of timing. That is not by chance. Speech operates at a much faster scale than other sounds, including music. Consider this: *allegro* is a musical tempo in the range of 120–170 beats per minute (bpm). For the sake of easy math and avoiding fractions, let's consider an *allegro* piece of music at 150 bpm. That equates to two and a half beats—quarter notes—per second. So each quarter note is a leisurely 400 milliseconds (ms, thousandths of a second) in duration, an eighth note is 200 ms, and a sixteenth note is 100 ms. "The Flight of the Bumblebee," at an even faster *presto* tempo, famously capitalizes on the fact that it generally takes a full 100 ms to tell two notes apart. By explicitly making the sixteenth notes in the main-theme melody rush past at around 80–85 ms each, Rimsky-Korsakov turned the notes into a beelike buzz. Speech is a different animal, however. Consonants in speech are *routinely* that fast or faster, on the order of 20 to 40 ms. And we can produce speech jam-packed with consonants almost indefinitely. "Flight of the Bumblebee" is mercifully short, to the relief of any musician who has played it.

More Sound Ingredients

Intensity is a measurement of the magnitude of pressure changes in air we perceive as loudness—how much air did the guitar string in figure 1.1 move and how tall are the waves that it made in figure 1.3? The absolute size of the changes in air pressure that produce sounds is tiny. Yet the *range* of air pressure changes that spans the quietest to the loudest sounds we experience is enormous—a whopping ten *trillion*-fold difference in physical air pressure. Thus, to shoehorn our loudness perceptions onto a set of reasonable numbers, we use a logarithmic conversion to turn the quantity of air moved into the familiar unit of sound intensity, the decibel (dB). That ten trillion-fold span can be expressed as the difference between 0 dB, the threshold of hearing—down at the limit of the most sensitive microphones—and 140 dB, the loudest sound we can tolerate.

Amplitude and frequency modulation—AM and FM—are terms you likely think about only when you turn on your radio. But AM and FM are extremely important to our auditory landscapes and especially so for speech. AM is a fluctuation of sound intensity (amplitude)—loud-soft-loud-soft. Many car alarms pulse in this loud-soft fashion. The vibration of our vocal folds as they open and close is amplitude modulating what we're saying at our voice pitch, the fundamental frequency. Figure 1.4 shows a basic form of AM—the same signal is being amplitude-modulated at two different rates.

FM denotes a change in frequency over time. As our speech morphs from consonant to vowel and back again, concentrated bands of acoustic energy sweep up and down. This is frequency modulation, the *FM sweeps* of figure 1.9.

Another sound ingredient that warrants mention is *phase*. At the beginning of this chapter, we arbitrarily showed the pressure of the air molecules to the right of the guitar string. The air molecules on the left in figure 1.1, which are not shown, spread out when the ones on the right are compressed, and vice versa. At any given moment, the movement of a guitar string is simultaneously compressing and dilating air molecules in the vicinity. Two people sitting on opposite sides of the guitar will hear music that, signal- and pressure-wise, is 180 degrees out of phase. A plot of the waveforms they are hearing would be reversed top-to-bottom. Depending on where you are sitting, the sound from the guitar will arrive at your ear at a different time, or phase. These different phases of the sound are important for sound localization, and phase additions and cancellations play a role in distinguishing sounds in reverberant (echoey) and noisy spaces.

Finally, there is *filtering*. Filtering is simply the selective reduction or enhancement of certain frequencies in a sound signal. We experience filtering a million times a day, both intentional and unintentional. Your favorite song sounds different whether you listen to it on your home stereo system, in the car, through your computer speakers, through earbuds, or through the speaker of your cell phone. Each sound reproduction system has its own filters, either carefully crafted by an acoustical engineer or simply as an unintended consequence of tradeoffs in size, cost of production, or other expedients. The voices of you and your friend talking sound different as you walk from the street into the coffee shop. The filtering caused by the hard surfaces of the walls, floor, and tub is why we enjoy singing in the shower. By the same token, Gothic cathedrals rely on shaped stone surfaces that create multiple reflections of the higher frequencies, giving these spaces distinctive acoustical properties for music and speech. Try listening to the speaker of your cell phone as you walk in and out of different rooms. Leaving filtering by external spaces aside, we deliberately filter the sounds we make with our mouth, tongue, and lips as sound is routed through and around them to achieve the words required to get the message across.

Signals Outside and Inside the Head: Ingredients

Our brain makes sense of signals outside the head—sound—with the signals inside the head—the electricity of neural impulses.

All scientists pick strategies for their inquiries. Some utilize surveys. Others use gene expression. Still others use blood biomarkers. My chosen milieu is signals. I find that signals—whether they are outside the head or inside the head—are reassuring because they are tangible, in some ways more than ephemeral sound itself. They can be measured with confidence, and there are widely accepted and powerful ways to visualize and analyze them. I find the remarkable similarity between the signals outside and inside the head most satisfying. This is a thing of beauty. It is a wonder that this happens. This tangibility gives me something I can hang my hat on, something to ground me when I research big ideas like the impact of music training on the sound mind, beat keeping's role in literacy, or how concussions can affect sound processing. I rely on signals to guide my thinking and to tell me the Truth.

The ingredients of sound are key to understanding why each person hears sound in the world differently, and how an individual's experience of sound can change for better or for worse as our sound mind is braided with how we sense, think, feel, move.

As a neuroscientist, I am able to bring this tangibility to my study of sound and its processing in the brain. I can study the processing of pitch, timing, and timbre in isolation and as an aural whole in the pursuit of figuring out what goes right and what goes wrong in people who are expert listeners and in those who have difficulties. Sound ingredients are separable in terms of how we process them and turn them into our perceptions. For example, there are people who have difficulty distinguishing pitches but have no problem with timbral qualities of sound or vice versa. Others have difficulty only in timing. Musicians and bilinguals alike are listening experts, but their prowess with signals operates on different sound ingredients.

Now let's see what happens when sound waves outside the head create brain waves inside the head—when the movement of that guitar string makes its way into the ear canal.

2

Signals Inside the Head

Ingredients Outside and Inside

At some point deep in our evolutionary past, natural selection steered us toward the ability to detect with our ears the pressure changes caused by tiny movements of air molecules. So we developed a series of body parts that, in a few fascinating steps, turn the air movement caused by a vibrating guitar string or a spoken word into the amalgam of ingredients—pitches, timbres, and timings—we perceive as a guitar or a voice.

To *transduce* means to change from one state to another. The currency of the nervous system is electricity. If we want to make sense of sound and to act on it, we need a way to transduce air movement into brain electricity. How do we do that? We start in the ear and follow an elegant sequence of events involving physical movement of bones, perturbations in fluids, and the release of chemicals. Then the signal moves to the brain, taking the electrical impulses the ear created and processing them further so that our sound minds can make the most of the sounds outside our heads.

I like to think of the brain's processing of sound as a mixing board. Like a sound engineer in a recording studio who slides the faders up and down to achieve a balance between the guitar and the vocals, the brain emphasizes some sound ingredients and deemphasizes others (figure 2.1).

Once transduction has been accomplished and we are working in the comfortable environment of electrical signals, we can visualize them in the same time, frequency (spectrum), and frequency-over-time (spectrogram) plots we use to think about sound. Like signals outside the head, signals inside the head entail the same *ingredients* like frequency, timing, and harmonics processed distinctly like dials or faders on a mixing board. The faders are set differently in every brain, due to experience, expertise, deprivation, or decline. Every sound mind is unique.

Figure 2.1
The sound mind processes sound ingredients to get the most out of them.

Upstream and Downstream

The sound mind is vast. When we hear, electrical signals course throughout the brain, moving upstream *and* downstream, interacting with our other senses, how we move, how we think, and how we feel. This entire brain network enables us to *make sense of sound*—to create meaning from our sonic world (figure 2.2).

Efferent and *afferent* are adjectives describing direction of movement, namely "away from" or "toward," respectively. Away from or toward what? In blood circulation, the answer is the heart. Vessels that carry blood from the heart outward are efferent; those with blood flow toward the heart are afferent. There is also afferent and efferent flow in the lymphatic system, carrying lymph fluid toward and away from the lymph nodes. In the world of neuroscience, the brain is the node. The *afferent* system moves information from the ear toward the brain. The *efferent* system* moves information away from the brain, back *toward the ear,* and, in so doing, is a cornerstone of how we learn—how we construct our sonic reality and become our sonic selves.

Moving Upstream (Afferent)

The upstream journey of electrical signals from the ear through the brain is the main course of this chapter. A Google image search for "auditory pathway" returns pictures reinforcing the classical view of a hierarchy of hearing—a preponderance of block diagrams with one-way, upward arrows from ear-to-brain as in figure 2.3. This is not *wrong*—indeed, the auditory brainstem lies between the auditory

---

*The dictionary tells us these two words are pronounced with short vowels. But in practice, to avoid mishearings, scientists often take pains to overaccentuate the initial syllables and pronounce them with long vowels: AYE-ferrent and EEE-ferrent.

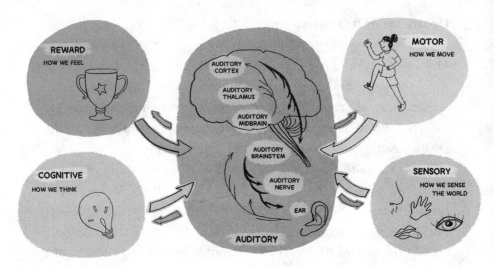

Figure 2.2
The auditory pathway has bidirectional connections among its own structures and
to brain areas responsible for sensing, thinking, feeling, and moving.

nerve and the auditory midbrain. The thalamus is situated between
midbrain and cortex. But it is only one part of the full picture. There is
absolutely a two-way flow of information, and it usually does not flow
hierarchically. Yet while I object to a hierarchical view of the auditory
system, I concede that a one-way model has its place in an overview.
Here, we will follow the upward arrows of afferent (toward-the-brain)
processing. We will end the chapter with a sketch of downstream
influences as a preamble to their more extensive treatment later on.

The Ear
The outer ear   The outer ear, the part we can see, funnels sound to
the ear canal toward the middle ear.

The middle ear   When a pressure wave caused by the movement
of air finds its way into the ear, past the outer ear and through the
ear canal, it hits the eardrum, also called the tympanic membrane.
Unlike some common terms for anatomical structures, like funny
bone and belly button, "drum" is an accurate description of this

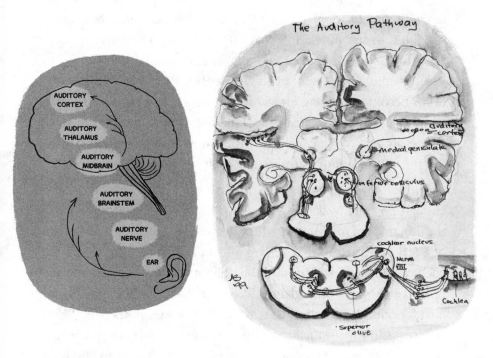

Figure 2.3
The auditory pathway through the brain corresponding to the block diagram on the left. Water-color rendition by Arnold Starr, MD, a pioneer at applying the brain's response to sound to assess neurological health. Reproduced with permission; photo by Tom Lamb.

threshold to the middle ear. Like the head or skin of a drum, the eardrum is a membrane that stretches when struck by sonic pressure. When this tiny drumhead moves, it pushes on the first of the three tiniest bones in our body—the ossicles*—which in turn push on the next and then the final ossicle—the stapes. The stapes then bumps into another anatomical drum, the even-tinier oval window, the gateway to the inner ear. Why do we need two "drums" separated by three bones? Because there is fluid on the other side of the oval window inside the inner ear. The movement of air alone is not

---

*Not only the smallest but the only bones that do not grow after birth.

forceful enough to push directly against the oval window, because the fluid on the other side is too dense to be moved by air itself. This three-bone linkage acts as a lever and amplifies the power of the movement by a factor of about twenty.* The tiny tap on the eardrum becomes a strong knock, forceful enough to nudge the oval window. Note that we are still in the mechanical-movement stage of the process. We have shifted from moving air to moving fluid. But the all-important transduction to electricity is still to come.

The inner ear (cochlea)   The tiny stapes bone now moves with enough pressure to displace the oval window and thus the fluid on the other side of it. This fluid whooshes past the hair cells of the organ of Corti—a structure that runs the length of the coiled-like-a-snail cochlea and just misses out on the title for the smallest organ in the body (curse you, pineal gland). See figure 2.4. All along the cochlea are hair cells; this is where the transduction magic takes place.[†] Hair cells come in rows—one inner and three outer—and each is crowned by strands of even tinier stereocilia, which gently bob in the fluid like the hair of a swimmer under water. The hair cells are sandwiched between the basilar and tectorial membranes whose names have an architectural basis—*basilar* is etymologically related to the word "basement" while *tectorial* comes from the Latin *tectum*, "roof." The hair cells are planted in the basement and the stereocilia are not free-floating; rather, their tips are attached to the roof. When fluid is set

---

*The middle ear deploys two mechanical engineering principles to magnify pressure between the eardrum and the oval window. The first is the lever principle: together, the three ossicles form a seesaw with the fulcrum nearer the oval window end. So a small amount of pressure on the eardrum is translated to higher pressure on the oval window, just as a small child can lever an adult into the air on a seesaw if the fulcrum is in the right place. The second is due to the size difference between the eardrum and the oval window—the latter is much smaller. Pressure is equal to force divided by area ($p = F/A$). The force does not change between the eardrum and the oval window, so the smaller area of the oval window (in the denominator) results in higher pressure.

[†]My first exposure to hearing science was a job peering through a phase-contrast microscope counting cochlear hair cells. Often doing this in the quiet of night, I found these tiny, elegant structures captivating.

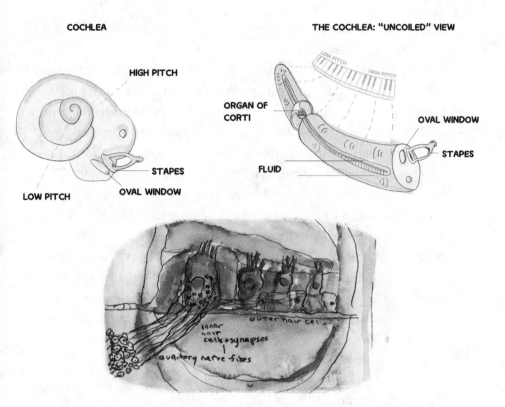

Figure 2.4

Top: the cochlea, coiled and uncoiled. The base of the snail-shaped cochlea, where the stapes meets the oval window, is tuned to high-frequency sounds. The apex, the center of the coil, prefers low frequencies. The "unrolled" cochlea on the right depicts this schematically with a keyboard and also includes a cross-section showing the organ of Corti within. Bottom: the organ of Corti. Here we can see one inner and three outer hair cells (sandwiched between the tectorial and basilar membranes) and their connection to the auditory nerve. Reproduced with permission from Arnold Starr. Photo by Tom Lamb.

in motion by a tap on the oval window, some of the hair cells bob up and down, causing the stereocilia to tug on the tectorial membrane. This tugging motion effectively "opens up" the inner hair cell so that electrically charged chemicals, specifically calcium and potassium ions, are able to rush into it. These ions set off a chain reaction that culminates in a release of neurotransmitters into a *synapse*, a junction between the hair cell and the auditory nerve, resulting in a sudden change in electrical voltage in the auditory nerve. Finally, we have achieved transduction. Air movement on the outside of the head has turned into electricity inside the head.

A given hair cell within the cochlea (of the ~30,000, total) does not bob around indiscriminately to every sound. The basilar membrane, where the hair cells are planted, has neither a consistent width nor stiffness along its length. The end closest to the oval window is the narrowest and stiffest, and as we move away from the base toward the apex, it gets systematically wider and floppier (like a ponytail). These physical differences bias hair cells on the narrow and stiff end to react to the highest-frequency (pitch) sounds. As sounds get lower and lower in frequency, they most effectively perturb hair cells closer and closer to the floppier apex. This systematic arrangement is called *tonotopy* (think "tonal topography"). First emerging in the cochlea, a *tonotopic map*, like a tiny piano keyboard, appears again and again throughout the auditory system, from cochlea to cortex. Maps in the brain are a fundamental organizing principle that spans our senses.

The Hearing Brain

We hear with our brains. One of my favorite embodiments of this statement comes from Robin Wallace's book *Hearing Beethoven*.[1] How did Beethoven compose some of his masterpieces after he had lost his hearing? The same as he always did:

> He improvised. He sketched. He revised. There was no dramatic change, no before deafness and after deafness. There was only an ongoing refinement of his relationship with the piano. Rather than envisioning Beethoven as a bird without wings or a fish out of water, we might think of him as a pilot flying safely without working navigational instruments, but with a deep bodily knowledge of how to steer an aircraft.

After the outer, middle, and inner ear have done their parts, there is still a long way to go before we can call it "hearing"; that is, before we can make sense of sound. Enter the brain. There are many way stations on our tour along the auditory pathway.

The word "brain" often connotes the cerebral cortex—the deeply grooved, multilobed, right and left hemispheres of the outer shell. I believe we should be equally attentive to the less notorious regions the cortex sits on. Between the auditory nerve and cortex are the cochlear nucleus, superior olivary complex (brainstem), the inferior colliculus (midbrain), and the medial geniculate (thalamus). Our transduced electrical signals traverse these structures on their journey through the brain. This trip involves more such structures than are found in any other sensory system.

Let's take a journey from the auditory nerve to the auditory cortex. Sound processing is *transformed* as it moves through the auditory brain. By recording simultaneously from neurons in the midbrain, thalamus, and cortex, Brainvolts alumna Jenna Cunningham showed us firsthand that neural responses are distinct along the auditory pathway. Her experiments made it plain to see that the response to the same sound differs from structure to structure.[2]

Auditory nerve   The auditory nerve is a bundle of fibers, approximately 30,000 per ear, tuned to particular frequencies depending on where they interface with the basilar membrane of the cochlea. The tonotopy (little piano) first emerging in the cochlea is next seen in the auditory nerve. Sound frequency is coded by *where* along the tonotopic map a neuron is situated. Tonotopic maps proliferate as we move toward the brain.

As we move from ear to brain, there is another organizing principle: The speed limit of neural firing decreases as the brainward ladder is climbed.* That is, how fast a given neuron can synchronize to sound in real time systematically declines from ear to brain. Auditory nerve fibers are the fastest.

---

*A neuron that fires to each cycle of a sound is *phaselocking,* another way the sound mind keeps track of the frequencies a sound is made up of. Remember that the higher the frequency of a sound, the faster it will complete one cycle, and so a neuron must fire at increasingly fast rates as sound frequency increases.

Cochlear nucleus Once transduction to an electrical signal has taken place at the junction of the cochlea and the auditory nerve, the first structure encountered on the path to the auditory cortex is the cochlear nucleus. It has many cell types with some pretty great names (bushy cells, cartwheel cells, octopus cells!)[3] and response characteristics[4] to get their jobs done. I'm showing you what these cells look like in figure 2.5 just because I think they're beautiful.[5]

As we ascend the ear-to-brain chain, a neuron's response to sound becomes increasingly specialized through the principle of *inhibition*. Neurons are not completely inactive in the absence of sound; they fire spontaneously. The response to sound can include both excitation (above spontaneous rate) and inhibition (below spontaneous rate). When a sound at a given frequency is heard, the firing rate of neurons tuned to that frequency increases above the spontaneous rate. Meanwhile, the firing rate of neurons tuned to nearby frequencies falls *below* the spontaneous rate. Inhibition helps certain sound ingredients "stand out," thereby increasing precision and tuning.

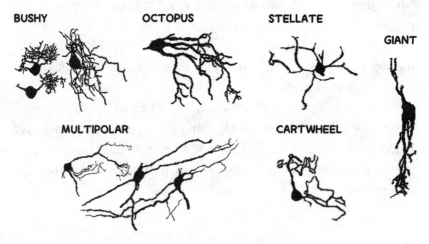

BUSHY   OCTOPUS   STELLATE   GIANT

MULTIPOLAR   CARTWHEEL

Figure 2.5
Cell types found in the cochlear nucleus. Adapted by permission from Springer Nature, *The Mammalian Auditory Pathway: Neuroanatomy*.

A specialty of the cochlear nucleus involves amplitude modulation (AM).[6] Cells here are specialized for certain AM frequencies. The pitch of our voice is determined by AM. When we speak, our voice is amplitude modulated as the vocal cords vibrate (open and close).

Once these refinements are achieved in the cochlear nucleus, neural impulses are passed along to the next structure in the chain, but the trip is longer this time, because for the first time, neural electricity from each ear is routed to *both* sides of the brain.

Superior olivary complex   The auditory system really shines when it comes to timing precision—leaving the visual system in the dust. The microsecond timing cues that exist in sound require microsecond precision in the brain. The superior olivary complex is where much of the timing wizardry happens, particularly as it pertains to binaural (*bi*-two; *aural*-ears) processing, sound location, and selectively picking out sounds of interest in an auditory scene.

Any sound that is not directly in front of us will arrive at the two ears with differences in timing and loudness. If a sound is coming from the left, it will arrive at the left ear a tiny fraction of a second earlier than the right ear. If the sound is just little bit off-center, the differences in timing between the ears can be as little as 1/100,000th of a second (10 microseconds). It also will be somewhat louder in the left ear than in the right, because it traveled a tiny bit less and because it wasn't blocked by the head. These timing and loudness differences between the ears are weighted differently depending on the sound's frequency. A low-frequency sound, due to its long wavelength, works its way around the head with little loss of loudness. However, the time of arrival differs enough for us to detect these microsecond differences. In contrast, a high-frequency sound *is* blocked by the head so there is a detectable loudness difference at the two ears. Because each ear routes information to both the left and right superior olivary complex, comparisons in timing and loudness are possible.[7] This helps us determine where in space a sound came from. Okay brain, do some math, figure out

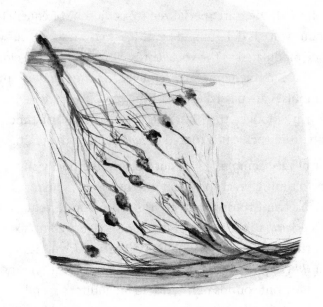

Figure 2.6
Signals from both ears converge in the superior olivary complex where their relative timings and intensities are analyzed. Reproduced with permission from Arnold Starr. Photo by Tom Lamb.

what position "in the world" would result in the particular timing and loudness difference my two ears just experienced. In addition to pegging a sound's location in space, this ability helps us group sounds together into an "auditory object" such as a companion's voice, so we can pay attention to it despite competing sounds in the soundscape. If your friend is sitting to your left in a noisy restaurant, it is tremendously helpful to be able to ignore the woman with a similar-sounding voice coming from the table on the right. The binaural processing that makes understanding in this circumstance possible is courtesy of the superior olivary complex.

Auditory midbrain—inferior colliculus The next stop in the afferent chain is the bump of the inferior colliculus (from Latin, "lower hill"), located in the midbrain. Inferior describes its position relative to the "upper hill," the superior colliculus; it is a statement of neither size (it is the largest auditory subcortical structure) nor importance (it is in the middle of the action). Because this metabolically active (energy-hungry) structure is both a hub of afferent auditory processing and a major crossroads of efferent, multisensory, and nonsensory neural activity, the functioning of the fittingly named *mid*brain is of critical interest to an auditory neuroscientist as a proxy for auditory function as a whole.

All of the signals from the auditory structures we have mentioned so far converge on the auditory midbrain from both ears, as do inputs from other parts of the brain. Calculations related to tuning selectivity, sound localization, and creating "auditory objects" must therefore be maintained in the midbrain.[8] Because of its central role as an assembler of auditory processing and a meeting site of brain signals from many sources, the auditory midbrain plays a crucial role in making sense of sound.

Luckily, despite being located deep in the middle of the brain, the midbrain produces an electrical signal robust enough to be measured from the scalp. Much of Brainvolts' research has been devoted to measuring this midbrain electricity, in the form of the "frequency following response (FFR)," and using it as a point of departure to study the brain mechanisms underlying music, reading, autism, aging, and more.

Auditory thalamus—medial geniculate The final station on the path to the cortex is the medial geniculate nucleus (*genu*=knee in Latin, as in genuflect, named for its bent shape). It is located in the thalamus, parked next door to the lateral geniculate, which is the subcortical processing center of the visual system.

It is worth pausing for a moment to consider that the visual system has a great deal *less* subcortical processing than the auditory system. The optic nerve runs more or less directly from the retina to the thalamus. There is no visual analogue to the auditory processing stations cochlear nucleus or superior olivary complex or inferior colliculus. It goes retina—thalamus—cortex—boom!* Likewise it is olfactory receptor cells in the nose—olfactory bulb†—cortex—boom![9] It is also worth noting that the various stations of the hearing brain—auditory nerve, cochlear nucleus, superior olivary complex, inferior colliculus, medial geniculate—are each made of a number of substations. The auditory subcortical system is uncommonly rich.

The thalamus relays input from the auditory midbrain to the auditory cortex, codes durations of sounds, accomplishes additional processing of complex sounds, and integrates considerable information from disparate brain regions. It regulates consciousness—alertness, arousal, and awareness. Think of the thalamus as a searchlight (it's even shaped like a light bulb) on the lookout for activity throughout the brain.

Auditory cortex   The auditory cortex is located, fittingly, above the ears in the temporal lobes, one on each side. The auditory cortex, containing multiple tonotopic maps, represents a final step of afferent processing. Binaural processing is refined, with bands of specialized neurons responding optimally depending on whether one or both ears receive the signal.[10] The auditory cortex contributes to the interpretation of harmonics,[11] consonance and dissonance,[12] and AM and FM signals.[13] The auditory cortex is a master at detecting sound patterns.[14] Neurons here often respond selectively to sound onsets,[15] thereby telling us when sounds start and stop. There is a wide range

---

*Despite it being a flight with fewer layovers, the visual journey takes longer. Whereas the initial transduction from sound pressure waves to electricity in the brain is essentially a single step, the retina must first transduce light to a chemical, which in turn triggers the subsequent transduction to electricity. Once this front-end bottleneck is overcome, auditory and visual neural signals move at the same pace.

†Olfaction is the only sensory system that bypasses the thalamus.

of specificity in cortical neurons; some are tuned for particular frequencies in a tonotopic fashion, but most are primed to respond only to certain combinations of sound ingredients (for example, FM sweeps that occur when consonants transition into vowels).[16] All in all, the flexible auditory cortex helps us pick out relevant elements from the ongoing soundscape to form discrete auditory scenes.[17]

Aside from these wide-ranging and specialized sound processing duties, the auditory cortex is responsible for the actual *recognition* of sound, in the tree-falling-in-the-woods sense. An intact ear and a full complement of functioning subcortical nuclei, dutifully firing electrical impulses in response to sound, will not result in *what we perceive as sound* without the auditory cortex.[18]

The lateralized sound mind The concept of left and right brain is one most of us are familiar with. Specialized operations administered by the left or right side of the brain is an evolutionarily ancient feature of the nervous system.[19]

From the point of view of the sound mind, the processing of *sound ingredients* filters into the left and right domains. For example, in speech, the fundamental frequency (pitch) is right-brain preferred while timing and harmonics, both phonetic cues, are left-brain preferred.[20] Sound, and the brain's response to it, unfolds over multiple time scales from microseconds to seconds. The processing of these timescales also aligns to one side of the brain or the other. Speech and music are processed in *both* cerebral hemispheres, but in different ways.[21] Distinctions in sound processing (pitch/timbre; long/short time scales) also exist subcortically.[22] Thus, the fundamental principle of brain laterality is found throughout the auditory pathway—another testament to the distributed, integrated, and reverberating nature of the sound mind.

The magic of hearing depends on the entire processing system working together, as we will learn in just a moment when we meet Peggy, David, and Susan.

Failing to Make Sense of Sound—When Signals Inside the Head Hit a Roadblock

At Brainvolts, we have the opportunity to see firsthand the real-life consequences of problems in a particular stage of processing. Individuals with unusual hearing problems often find their way to us.

Meet a young woman we shall call Peggy, who has auditory *cortex* damage known as "cortical deafness." She had undergone aggressive treatment for cancer that saved her life but damaged her auditory cortex on both sides. Peggy's ears and subcortical structures work well. But due to the damaged cortex, Peggy is *aware* of sound but cannot *understand* it.

David on the other hand, is a child who has a problem with *subcortical* processing of sound. David's parents and teachers knew something was wrong with his hearing. David had extraordinary difficulty hearing anything in noisy places like a classroom. He was failing to turn in his school assignments simply because he didn't hear them being assigned. He also responded inconsistently to sound at home, making his parents wonder if he had a hearing loss. Yet his ears checked out just fine. He nailed the test where he had to indicate he heard the beeps at all the different pitches even when they were very quiet. It turned out to be a lack of synchrony in neural firing in the *subcortical* structures of the brain. Neural activity was making its way from the ear to each way station up to and including the auditory cortex, but not in a synchronized manner. The timing was all wrong.

David's collection of symptoms is a now well-known condition called auditory neuropathy.[23] Its hallmark is having a terrible time hearing if there is the slightest amount of background noise—truly deaf in noise. In a quiet setting, there is often no problem understanding what is said. Unlike with cortical deafness, people with auditory neuropathy are often *not aware* of sound in the first place. One young woman with auditory neuropathy—let's call her Susan—whom we have been following at Brainvolts for twenty-plus years, resorted to wearing earphones at work to make her coworkers think she was listening to music even when she was not. The earphones prompted her colleagues to tap her on the shoulder when they wanted her attention because she was unaware they were calling her name. Now her little daughter alerts her mom when someone is at the door or the phone is ringing.

People like Susan, David, and Peggy teach us. They tell us we need the auditory cortex to understand sound. And they tell us the subcortical auditory system, and the exquisite and fast and synchronous and consistent neural firing it is known for, is necessary for sound awareness, and to maintain signal clarity for hearing in noise—for navigating the auditory scene. David and Susan help us realize why hearing is our fastest sense, and how it relies on *exquisite synchronous timing*. The slightest amount of sluggishness has serious repercussions. When these people come to Brainvolts, they are seeking answers, and in some cases we are able to see something in the way their brains react to sound to reassure them in a "well, no wonder you're having difficulty" sense. But really, they are teaching all of us what is possible. By showing us what can go wrong with our sound minds, they show us what is responsible for our hearing success when everything goes right.

The Ear-to-Brain Transformation—Questions and Answers
What is both exciting and humbling is how much we *don't* know. For example, it is not unusual for a given structure to have multiple side-by-side frequency (tonotopic) maps.[24] Why do the maps proliferate? How do they differ in function? To pick another example, both the superior olivary complex and the auditory cortex play pivotal roles in binaural processing, but we don't know much about the distinctive role each structure plays. The auditory midbrain gives us yet another puzzle. The input from way stations such as the cochlear nucleus and the superior olive all converge onto the auditory midbrain. One might think that after those structures had performed their distinct tasks, their outputs would not be dumped

back together. But they are. Why is the hearing subcortical network so much more massive and intricate compared to the other senses? I am certain there are elegant explanations awaiting discovery.

What we *do* know are principles about the transformation of sound as it traverses the afferent stream from the ear to the auditory cortex. Neural information is not simply inherited unchanged as it moves along the auditory pathway. Rather, neurons exhibit increasingly diverse firing patterns and become more selective about the sounds they respond to. Neurons become progressively more "interested" in when sounds stop and start. Inhibition, the suppression of the firing of certain neurons so that sound processing becomes more focused, becomes more common. The ability of neurons to change with experience also increases. These principles (diverse neuron firing patterns, inhibition, selectivity to certain sounds, changes with learning) contribute to the increased specialization we see as we move from the auditory nerve to the cortex. At the same time, as we move further along the chain, auditory centers increasingly interface with each other, other sensory systems, the motor system, what we know, and our feelings about sound.[25]

Another useful principle is the speed with which neurons synchronize to sound is fastest closest to the ear and progressively slower as we climb to the cortex. If a sound is repeated at a fast rate (rat-a-tat-tat), at thirty times per second, subcortical neurons can keep up with it no problem, but cortical neurons can only keep up with a much slower rate. Likewise, subcortical neurons can keep up with frequencies as fast as, say, 2,000 Hz. Cortical neurons can manage only about 100 Hz. The information is not lost as it ascends the pathway, but how it is encoded changes—the timescale of integration is longer higher up. *The microsecond timing precision that exemplifies auditory processing belongs to the subcortical realm,* including the fast calculation of timing differences between ears that lets us localize and identify sounds in space. The subcortical structures are the timing experts of the brain. On the flipside, there is a corresponding *ability to integrate auditory scenes over longer periods of time in the cortex*—a necessity as we navigate sentences and musical phrases.

In summary, subcortical and cortical networks work together to process sound. From a functional standpoint, the subcortical system enables us to hear signals in complex soundscapes, making it possible to hear our friend's voice in a noisy room. It is also essential to sound awareness in the first place. The cortex is essential for deriving meaning from sound, our ability to understand the words our friend said to us.

Moving Downstream (Efferent)

Recognition of the prominent role of the efferent system in how we sense the world is relatively recent. The auditory efferent system has a substantial brain-to-ear network that establishes back-channel communication alongside the afferent ear-to-brain connections. The efferent connections are more numerous than the afferent and are less like a train arriving at each station along the pathway. In short, everything talks to everything else. But why? The extent of efferent connectivity increases with evolutionary advancement,[26] and its dominance in humans and other highly evolved species plays a role in our mental flexibility and propensity to learn. The efferent system selectively emphasizes the sounds we learn are important.[27] I am using "efferent" expansively in this chapter, referring not only to the movement of information *within* the auditory system, but the movement of information to the auditory system from nonauditory brain centers.

What we hear is guided by downstream processes.[28] An implicit perception of a sound begins with a broad gist of that sound. Then, feedback from the auditory cortex, along with input from the cognitive, motor, and reward centers, triggers a scrutiny of the important details—along with a pruning of the unimportant—to arrive at a detailed perception of the sound. That is, the messages carried by our afferent system are informed, via the efferent system, by our past experiences with our life in sound. Our sound mind formats the reality of the signals outside our heads we perceive as sound. Each of the auditory way stations—auditory nerve, cochlear nucleus, superior olive, etc.—communicates with each other as well as with our other senses, how we move, what we know, and how we feel. It is precisely this interaction of downstream and upstream influences that allows learning to occur and sculpts our sound mind.

Hearing Engages Our Other Senses

Seeing influences hearing and vice versa. The gesture a percussionist uses to hit a marimba (an instrument related to a xylophone) influences the perceived length of the note. When a video of a percussionist playing a long note is accompanied by an audio clip of a short note, research subjects *hear* a long note.[29] Similarly, the judgment of vibrato in stringed instruments is affected by what we see. Vibrato is the slight warbling of pitch that comes about by a back-and-forth rolling of the fingertip on the string as the bow is drawn across the instrument. The amount of vibrato perceived in a violin note is influenced by whether the rolling finger movements that generate vibrato are seen, as compared to hearing that note without seeing it.[30] Even the distinction between a plucked vs. bowed cello string is blurred if a video showing a musician plucking accompanies the sound of bowing and vice versa.[31] A famous audiovisual interaction involving speech is the McGurk effect.[32] An audio clip such as "ba" will sound like "fa" if it is dubbed onto a video of a person producing the mouth motions required for "fa." The sight of the front teeth touching the lower lip that an "f" requires suggests an "f" (or sometimes a "v") is being produced. Visual priming tricks our brain into *hearing* "fa." Touch and smell also influence how we hear.

Hearing Engages How We Move

"What did you do to the piano? It's so much *easier to play.*" My piano teacher Salvatore Spina, who is also a piano tuner, says he often hears this from his clients after a tuning. The piano seems easier to play—as in requiring less physical effort. I suspect it has to do with increased feelings of relaxation. Listening to the dissonant sounds of an out-of-tune piano puts you on edge and tenses your muscles. A pianist who plays a well-tuned piano is a calm pianist. That's my guess, anyway, based on what we know about the communication that takes place between the auditory and motor systems.

There are rampant connections between hearing and moving. Hearing and movement have a common evolutionary origin. The ear arose from organs designed to perceive gravity and an organism's place in space with the goal of achieving movement. Merely listening to speech (without moving) activates the motor cortex as well as our own speaking muscles. Just listening to rhythm patterns[33] or piano melodies[34] activates the brain's motor system, particularly in musicians. The converse is also true—pianists looking at someone playing the piano without hearing it or people engaged in silent lipreading have active auditory centers.[35] Moreover, the movements that musicians make as they play influence the listener's perception of such things as the emotional impact or tension in the piece of music, even at an automatic physiological level.[36]

Mirror neurons respond whether you personally perform a movement or see or hear someone else perform that movement (figure 2.7).[37] These neurons help us figure out others' intentions and emotions from observing their actions. Mirror neurons may contribute to our feelings of empathy and to language learning. Deficits in the mirror neuron system have been linked to autism and may underlie why it can be difficult for an individual with autism to view the world from another's perspective, although this interpretation is controversial.[38]

Figure 2.7
Mirror neurons respond similarly whether performing an action or watching the same action performed by another.

Hearing Engages What We Know

One of my favorite demonstrations in my Biological Foundations of Speech and Music class involves a sound clip of a sentence that has been highly processed so it sounds like nothing more than a few seconds of garbled static. Think Darth Vader with a toothache doing a Cookie Monster impersonation during a thunderstorm. I play it a couple times and ask the class to raise a hand if they know what it is. Inevitably no hands go up—no one can even tell it's speech. Then I play the ungarbled version of the sentence. When I play the garbled one again, lightbulbs go off all over the lecture hall. Suddenly that garbled mess is completely understandable to every student. Everyone is amazed at how obvious (in retrospect) the garbled sentence was and can't believe it was ever challenging. What we know has an enormous influence on what we hear.

Hearing Engages How We Feel

"It's so good to hear the sound of your voice!" This comes from the sound-to-feeling connections we have made throughout our lives with the people we care about. The *limbic* or *reward system*, responsible for feelings of emotion, motivation, and reward, involves an array of structures in the cortex, brainstem, thalamus, and cerebellum. Parts of this system are among the most evolutionarily ancient in the brain. This is why sound is such a strong portal to our memories. Survival depends on remembering what danger and food sound like.

Whether you're a person, a monkey, a bird, a turtle, an octopus, or a clam, the physiological changes that accompany our deepest-felt emotions appear to be the same. Hormones and neurotransmitters, the chemicals associated with desire, fear, love, joy, and sadness are similar across taxa. Nearly all animal species have hormones like estrogen, progesterone, testosterone, and corticosterone (a stress hormone).[39]

Dopamine released during eating or sex, is linked to feelings of pleasure, regardless of species. Its release is also implicated in drug addiction and reduced responsiveness to pain. It contributes to a feeling of fear when a sudden sound disturbs a late-night walk. The limbic system has privileged access to hearing centers via fast, low-resolution pathways. This is why we might have an immediate visceral reaction to that late-night sound before the analytical brain kicks in and realizes a moment later it is a harmless trash-can lid banging in the distance. This speed of processing can be attributed to the subcortical and subconscious essence of emotion.[40] The midbrain response to sound is influenced by serotonin, another neurotransmitter involved in cognition and reward.[41]

The limbic-reward system is at play when a mouse mother responds to the calls of her pups. When a mouse pup strays from the nest, he will call out. The social behavior of returning the pup to the nest causes a release of oxytocin—a hormone associated with mother-child bonding. That release impacts how the auditory cortex processes sound ingredients. The same pup call elicits a dramatically different response in the auditory brain of a mother mouse compared to a mouse who has never given birth.[42]

As much as vision, motion, thinking, and feeling impact our auditory neurons, one of the biggest modifiers of sound processing is . . . well, processing sound. The sounds of our lives—our sonic experiences—leave indelible imprints on the very neurons that perform the task of hearing and turning sound into meaning. We learn because our neurons change and vice versa. Doing something again and again eventually makes us experts at that task; we say we can "do it in our sleep." Sufficient experience with deriving meaning from certain sounds changes how the sound mind processes those sounds automatically—even while asleep. This is because *efferent* modulation has driven *afferent* change. Response properties of neurons throughout the auditory pathway are malleable, right to the cochlea itself. These changes in how neurons alter their firing with experience make each of us respond uniquely to sound, as we'll see next.

3

Learning: Merging Signals Outside the Head with the Signals Inside

The sounds of our lives shape our brain.

Salvatore Spina, my piano teacher and tuner, is a new grandpa. His daughter is a world-class horn player. Last week, he was holding his three-month old grandbaby while Ralph Vaughan Williams's *Pastoral Symphony* played in the background. The baby was sleeping soundly until the opening of the second movement, which starts with a slow, quiet, haunting horn. She opened her eyes wide and looked around, only to settle back to sleep when the horn was replaced by strings thirty seconds in. Auditory learning starts early.

As a rabbit assigns a meaning to a sound—that is, after he learns a particular sound is relevant to his health and well-being—the pattern of neural firing to that sound changes (figure 3.1).* It was really *something* to watch this happen in an individual neuron in the

---

*Probing a neuron with a microelectrode is challenging in the best of circumstances and these experiments required the rabbit learning a task . . . which could take a while. The lab was on a busy street. I often performed these experiments in the middle of the night so passing trucks were less likely to vibrate the electrode away from its cerebral prey.

Figure 3.1
Sound processing changes in individual neurons when the sound has relevance.

auditory cortex. I felt I had nudged open a previously locked door. It made a big impression on me to witness learning firsthand in the raw building blocks of the brain, the individual neurons.

So much of how we perceive the world is unconscious. After training, the rabbit did not *consciously* fire off his neurons more vigorously, in the same way that I, as an Italian speaker, do not consciously make my brain perk up at the sound of an Italian voice. In this chapter, I want to reveal biological insights about how our sound minds are the product of our experience, in ways we are generally unaware of.

Neural plasticity is the catchall term for changes in the brain that result from experience. If I had to boil my career's work down to a two-word phrase, I could certainly do worse than "neural plasticity," although I would regret not shoehorning "sound" in there somewhere. While I care deeply about sound processing principles—which neurons fire in response to which sounds—I am most interested in how those patterns of firing come about and, by extension, how those patterns are altered as we create meaning from our sonic world. If I had to encapsulate into one sentence—two words is a bit restrictive—what my career has taught me, it would be the chapter epigraph: "the sounds of our lives shape our brain."

How does this brain shaping come about? The efferent system looping between the cortex, the subcortex, and the ears fuels auditory learning. This brain-to-ear network increases in scope and complexity with evolution. It is even more extensive in its neural projections than the "toward the brain" afferent leg that gets all the lecture-hall publicity. The most sophisticated and flexible parts of the brain are in constant dialogue with our more hardwired structures, thanks to the efferent neural pathways. The message delivered downstream, away from the brain in the direction of the sensory receptors—the cochlea, the retina, etc.—is the secret sauce to learning.

The changes in air pressure we register as sound are transduced by the ear and launched as electricity into the afferent (toward the brain) processing stream. Depending on the sound ingredients (pitch, timing, timbre . . .), a certain population of neurons in the cochlear nucleus, the superior olive, and so on will fire. That same sound, occurring a moment later, will knock over that exact line of dominos. But, as I witnessed decades ago in my rabbit experiments, if that same sound takes on new meaning, over time it may recruit a different set of neurons or speed up the rate of firing or commandeer a new location within the tonotopic map. The sensory-cognitive-motor-reward interplay of the afferent and efferent team shapes the *default* neural processing of the sounds important to us. Because the meaning of sound has changed, downstream signals have forged a new default pattern of afferent activity. Biologically speaking, *learning* and *memory* have taken place. This default system provides a runway onto which incoming sounds land, a mechanism for us to sense what is important. Exactly how my sound mind responds today depends on my life experience with sound up to now.

Maps

The "pianos" that appear all along the auditory pathway represent regions of maximal sensitivity to particular pitches—tonotopic maps. We see similar neural mapping in the other senses (figure 3.2). The visual system has retinotopic visual maps—the specific parts of the brain activated depending on where in the visual field a seen object is located. Maps pervade the somatosensory (touch) and motor systems as well. Both have their respective orderly and systematic maps representing body parts. The ten fingers are mapped to a sizable slice of somatosensory cortex. Other body parts where

Figure 3.2
Sensory maps are not unique to the auditory system. The exquisitely precise fundamental organization of our hearing system has analogues in vision, touch, and movement.

touch is important, like the tongue and lips, also get large parcels of brain real estate; elbows, shoulders, and legs get smaller parcels. You can observe this uneven somatosensory geography yourself. Close your eyes while someone touches you lightly with either one or two sharp objects like toothpicks. On any of your fingers, you can feel two toothpicks touching you if they are separated by as little as 3 millimeters (mm). However, to detect two distinct pokes on your back or thigh, the toothpicks must be 30–50 mm apart. Anything closer feels like a single poke. The nearby motor cortex is similarly arranged, with lots of geography given to body parts that need to make fine, precise movements such as the hands, fingers, lips, and tongue.

Some of the earliest discoveries of sensory learning came from observing changes in sensory maps. After somatosensory and motor maps were discovered in the 1930s by Penfield, Woolsey, and others,[1] it was believed this one-to-one mapping between body parts and brain areas was evidence the brain was hardwired. This notion was upended by Michael Merzenich, who demonstrated that after a monkey repetitively performed a task using the same two digits of the hand, corresponding cortical regions expanded. Similarly, if a nerve in a monkey's hand is injured, that corresponding cortical region does not go silent. Rather, it is taken over by other hand areas.[2] That is, the cortical map corresponding to the pinky doesn't die away if the pinky is injured but is commandeered by the other fingers. Merzenich, who also made some of the initial discoveries about tonotopy in the auditory cortex in the 1970s,[3] later expanded our understanding of cortical maps with the discovery that multiple maps can live in overlapping harmony. Various sound ingredients are simultaneously mapped in the auditory cortex.[4] In addition to the pianos that code how high or low a sound pitch is, there are maps that code how loud or soft a sound is or map the position of sound in space. Work on auditory cortex plasticity demonstrated how flexible the auditory maps actually are.[5]

Cross-modality map plasticity is observed too. The visual cortex in blind people can be recruited by the auditory[6] and somatosensory[7] systems. Piano tuning is one of the classic "blind trades," due to the hypersensitivity to sound often present in blind people. Conversely, in deaf people, the auditory cortex is taken over by visual processing used for communicating in sign language.[8] This suggests a profound capability for neural reorganization—a property necessary for auditory learning.

The Owl Story

One of my favorite accounts of auditory learning involves the barn owl and some psychedelic eyeglasses. The barn owl is a nocturnal predator. As such, it does not have the luxury of the sun illuminating its prey. So owls rely on sound location cues to hunt. Their sound localization is about twice as good as ours—they can resolve a sound in space, in either the horizontal or vertical plane, down to about a 1-degree angle.[9] What does a 1-degree angle translate to? If I stand on the goal line of a football field with my arms outstretched and snap a finger, an owl on the opposite goal line, using sound alone, could figure out whether I snapped with my right or my left hand. Barn owls can locate sounds anywhere in their spatial field and, due to differences in ear height and orientation (one ear points down and the other points up!), can also achieve precise localization for elevation, something very difficult for humans.

Owls, like humans, use both timing and loudness differences between the ears to localize sound. The frequency of a sound determines which of these cues humans use for localization—we predominantly use loudness cues for higher-frequency sounds and timing cues for lower-frequency sounds. The owl uses *both* cues simultaneously for any given sound regardless of frequency. Timing differences between the ears are used to determine right/left location, and loudness differences are used to determine elevation.[10] In this way, owls have enough information to metaphorically plot any sound on a sheet of strigine graph paper.

So where do the psychedelic specs come in? The *spatial sound maps* that owls construct align with *spatial visual maps*. Neural integration between sight and sound keeps these maps in alignment. However, this alignment must be learned. A mouse squeak that arrives at the right ear slightly louder and slightly earlier than in the left ear does not have any spatial meaning until a young owl learns to associate that particular timing/loudness blend with lunch being served in the bramble to the upper right and not the weed patch to the lower left. In this way, an auditory spatial map formed in the auditory midbrain interfaces with a visual spatial map in the visual midbrain. With the help of efferent-mediated coordination and memory, these two midbrain maps, over the course of development and experience become aligned. Now, a squeak arriving at the owl's left ear at, say, fifty microseconds earlier than at the right prompts a lightning-fast head movement to the precise space in the left visual field that corresponds to that particular delay—about 20 degrees off-center—precisely where the unlucky mouse is.

Enter the neuroscientist.

It is possible to fit prisms onto an owl's eyes, like goggles (figure 3.3), so the spatial positioning of its visual world is shifted. Let's say our owl had learned that a sound with a particular between-ear timing difference meant the noise-making object is a bit to his left. But now that same sound, coming from the same auditory location, is associated with an object to the right, thanks to the distorting goggles. After wearing goggles for a few weeks the owl will create a new audiovisual space map so he now knows to turn his head to the *right* when that particular sound occurs. The hunt is a success, the mouse is in the owl's tummy, and we have ourselves a beautiful example of efferent-mediated learning.[11] Learning a new sound-vision map, motivated by hunting success or failure and

Figure 3.3
A bespectacled owl.

fueled by the efferent system, resulted in a change of the receptive properties of the auditory midbrain. (In case you are wondering, after the prism-induced reorganization is accomplished and the prisms are removed, the owls' spatial maps ultimately return to normal but not immediately.)

Does Auditory Learning Have an Age Limit?

We know the juvenile brain is primed for learning. It was once thought that spatial-map reorganization was restricted to young birds. No prism-induced shift in the spatial map was initially observed in older owls,[12] suggesting subcortical reorganization might be impossible after the sensitive period of youth. However, it turns out younger and older animals have different learning strategies. Taking a graded approach yielded different results. Instead of jumping immediately to a prism that shifted the visual map 23 degrees, as with the young owls, the adult owl was fitted with prisms that shifted the map by only 6 degrees. This shift led to successful learning and, with successive small shifts, the older owl eventually exhibited the same extent of map reorganization as was possible in the juveniles.[13] These owl discoveries, and many others like them, are a cause for optimism. *Learning is always possible, at any age, given the right approach and optimal setting.*

Of conspicuous significance, learning at *any* age was faster when owls lived in an enriched environment (a large aviary with more opportunities for stimulation, exploration, interaction with other owls), compared to being caged on their own.[14] We will revisit the power of enriched (and impoverished) environments on the sound mind again and again.

Learning Throughout the Hearing Brain

The owl story reveals key biological mechanisms underlying how we learn to perceive the world. It reveals the power of the efferent system as experience initiates a fundamental rewiring of the sound mind. It shows us there is a great deal of cross talk between the senses. It offers evidence that with the proper environment, neural reorganization is possible throughout life. It also has a special emphasis on the *timing* ingredient of sound, my personal favorite.

What happens *throughout the brain* as we form sound-to-meaning connections? Learning occurs in all parts of the auditory pathway: the cortex, the subcortex, the auditory nerve, and the ear itself. The efferent system is absolutely essential to the listening feats we take for granted.

Learning in Auditory Cortex

Any neuron in the auditory cortex, due to tonotopic mapping, has a preferred pitch (sound frequency) to which it responds best. Other frequencies might have little or no effect on its neural firing, and still other frequencies, usually those that closely flank the preferred frequency, might actually *inhibit* its firing.

Cortical maps are great at illustrating what happens in the brain when we learn to make sound-to-meaning connections. For example, in the ferret, an auditory-cortex neuron's preferred frequency might be determined to be, say, 8,000 Hz, with an inhibitory band

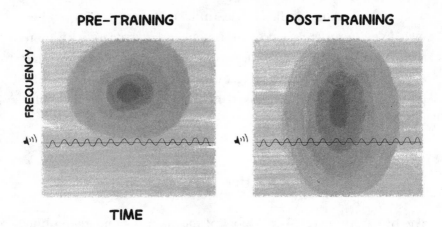

Figure 3.4
Neurons change with learning. Increasingly darker grays indicate increased neural firing. Prior to training (left plot), maximal activity is at a particular frequency, say 8,000 Hz. After the ferret learns a lower frequency of 6,000 Hz (wavy line) is important to listen to, the range of frequencies the neuron responds to expands to encompass the now-relevant frequency (right plot).

below this preferred frequency centered about 6,000 Hz. Then the ferret learns that a tone of 6,000 Hz signals something the ferret cares about—namely, a reward of some kind. After training, this same, formerly 8,000 Hz neuron expands its response to encompass 6,000 Hz, and fires less vigorously at its original preferred frequency (see figure 3.4). We're only looking at one neuron here, but other adjacent neurons (e.g. 7,000 Hz) also jump on board and respond to 6,000 Hz. There was a motivation-driven increase in the encoding of the newly relevant ingredient of sound.[15]

Learning in Auditory Subcortex

The ferret, like the owl and like us, uses differences in timing and loudness between the two ears to locate sound in space. And if sensory input is altered—by plugging one of the ears—sound localization ability, after taking an initial hit, can be relearned.[16]

Once auditory spatial maps are established (or reestablished by training), chemically deactivating the efferent connections from auditory cortex to the midbrain produces little immediate impact on localization ability—which has already been learned. However, *without efferent connectivity*, a new map cannot be established and *learning cannot occur.*[17] The reverse is true as well. Once a sound as lost its meaning for whatever reason, maps revert, but not without an intact efferent system.[18] An intact connection between auditory cortex and midbrain is essential for learning or unlearning to take place.

Another way of looking at how the efferent system changes brain tuning is by observing what happens by imitating efferent system activity. By electrically stimulating neurons in the auditory cortex directly, we can observe corresponding shifts in the neurons in the midbrain[19] and thalamus[20] that receive efferent connections from that cortical area. Midbrain and thalamus responses are sharpened either by recruiting additional neurons or by triggering inhibition.[21] This cortical influence extends beyond the midbrain to the cochlear nucleus,[22] several steps removed from the cortex.

Training-related changes to the human sound mind are instigated by top-down influence, in much the same way that owls and ferrets change their midbrain processing when new sound-to-meaning connections are learned. We can speculate, for example, that (mis)learning takes place in a child with many middle-ear infections. In these children, the hearing brain, like the ear-occluded ferrets, is fed a quieter signal (usually to only one ear). It is easy to imagine that auditory learning might be hampered during this sensitive period of development.[23] In subsequent chapters we'll explore auditory learning—how the sound mind changes for better and worse with our life in sound.

Learning in the Ear or, How Low Can We Go?

Is there any evidence the ear itself can alter how it works with training or other forms of efferent input? Before I answer, let me tell you something extraordinary. The ear itself can *make* sounds. (Imagine your eyeballs producing light!)

The inner ear (the cochlea) contains both inner and outer hair cells. The transduction from movement to electricity at the auditory nerve is the job of the inner hair cells. What role do the outer hair cells, which outnumber the inner hair cells three-to-one, play? These super receptors are on *the receiving end of efferent processing from the brain*. These highly complex structures can move by themselves[24]— and their movement is used to modify what the inner hair cells communicate to the brain. For example, more amplification for quiet sounds and less amplification for loud sounds increases the range of intensities we can hear. The ear is hearing the brain.

The movement of the outer hair cells creates audible sounds we can record with a tiny microphone in the ear canal. These sounds, with the rather dreary name of otoacoustic emissions (OAEs), can be induced by sound.[25] The OAEs occur only if the ear can "hear" at the frequencies of those sounds. This fact has revolutionized newborn hearing screening. It is now possible to determine in a few seconds whether the ear responds to a range of frequencies important for communication.

Knowing the ear creates sounds and does so by means of the part of the ear under efferent control reinforces the importance of the brain-to-ear system. It also provides a convenient window into how the brain communicates with the ear.

Here is how it is done. First, otoacoustic emissions are induced by playing sounds into, say, the right ear. The sounds that come back out represent the baseline cochlear activity. Then you repeat the process while simultaneously playing a loud noise to the *left* ear— the shhhhh of white noise will do nicely. The brain, upon receiving the news that the left ear is hearing a noise, exerts its influence on both ears, telling the outer hair cells in the cochlea to cool their jets and back off the amplification, for the purpose of protecting the ear from noise.[26] And this can be seen in the size of the OAE. The brain has thus exerted control all the way to the very first stage of sound processing.

There are several other ways the brain influences the ear. First, the size of the OAE is reduced when the auditory cortex is either damaged or electrically stimulated.[27] Second, if a person is instructed to pay attention to the sound rather than to simply relax, the size of the OAE is affected, again showing efferent control of the cochlea.[28] Third, musicians, with their lifetime of sound expertise, have distinctive OAEs and, presumably, a more finely tuned cochlea than nonmusicians.[29] Fourth, the size of an OAE is affected by whether you see a video of a person talking versus whether you hear the voice without a video.[30] Thus, sound processing in the brain and even our initial sound-sensing epithelium itself—the cochlea—has an entire efferent infrastructure that firmly controls it.

We Learn What We Pay Attention To

I am a guitarist; my husband, however, is a *Guitarist*. One day I was trying to figure out the lead in the Dire Straits' song "Sultans of Swing." There is a particular sequence of notes Mark Knopfler plays during his solo that I was struggling with. Dididi, dididi, dididi. I could not come close to plucking the string three times in such quick succession. My husband came by and said, "Nina, if you'd just *listen*, you'd hear he's pulling off the string with his left hand." (By changing the fingering on your fret hand, a pull-off enables you to play multiple notes during the course of a single right-hand pluck.) Notes that are pulled off have a distinct sound. It not only increases how fast you can play but shapes the timbre (harmonic ingredient) of the sound. After a while, I was able to hear it. I could recognize the difference in timbre. I expect my harmonics faders have moved up. But first I had to learn what to pay attention to. It was only after I made a concerted effort to attend to the harmonics a pull-off sequence made, that I really *heard* it. And only after time, effort, and explicit attention did hearing it become unconscious and automatic—my default response.

Attention belongs to the thinking dimension of our sensing-thinking-moving-feeling sound mind network. With attention, sensory maps are reorganized,[31] and the extent and long-term stability of the reorganization is directly related to the amount of effort expended in focusing attention.[32] Attention-driven learning is solidified by a corresponding release of dopamine[33]—a midbrain-generated neurotransmitter that mediates attention and is involved in reward and motivation.

Despite the billions of neurons in our brains and our sophisticated sensory systems, we are simply incapable of processing every image, every sound, every motion, every smell, and every last warm breeze we experience every single second. With this sheer volume of sensory input (estimated at ten or more megabits per second), we must prioritize our processing. We need to filter out what is unnecessary to focus on what is important *right now*, be it hunting, avoiding being hunted, listening to speech, reading, safely navigating the world, or enjoying a guitar passage. We accomplish this with attention. We have spent a lifetime learning what is important, and with this learning we have taught our brains which sounds, sights, and smells require our attention and which can be profitably ignored. David Strayer, University of Utah psychologist, said, "Attention is the holy grail. Everything that you're conscious of, everything you let in, everything you remember and you forget, depends on it."[34]

Attention in-the-Moment

A situation we face daily is hearing a friend in a noisy room with other people carrying on conversations. This is known as the cocktail party problem: we must harness auditory attention to tune in our friend's voice and tune out all the others.

The network in the brain that lets us tune in the wanted and tune out the unwanted is known as the *reticular activating system*. This joint cortical and subcortical system has direct access to the entire auditory pathway, allowing focused attention to change how neurons respond to sound.

Earlier, we saw how individual neurons in the auditory cortex shifted their previous tuning to encompass a newly important frequency when ferrets were taught to pay attention to the new frequency.[35] If the ferret learns to associate two different outcomes to two different tones—one they must ignore, one they must attend to—a neuron's tuning will doubly shift at *both* frequencies.[36] These shifts are not limited to frequency. If the learning task is designed such that another sound ingredient, say a timing cue, has meaning, the timing pattern of the neuron's response is altered suitably.[37]

Attention-induced shifts in the brain's tuning or timing of sound ingredients occur throughout the auditory pathway including the midbrain[38] and the auditory nerve.[39] The mechanism is probably a tempering of the amplification provided by the outer hair cells in the ear via the efferent system. This is why my husband can't hear me when he's reading a book.

A neural recording can be made while a person hears two simultaneous sentences but is instructed to pay attention to only one of the sentences. Brain responses to the combined sentences match the acoustics of the attended sentence better than the unattended one. In other words, focused attention to the first sentence suppresses the neural processing of the equally salient but contextually unimportant second sentence.[40] Context matters.

The sound mind works in concert with the limbic, cognitive, sensory, and motor systems to maximize our immediate listening goal. Our listening goal today might not match our goal tomorrow, and this flexibility is important. However, there are people who pay repeated attention to sound details. These sound experts give us insight into how repeated in-the-moment auditory attention can transform our sound minds into a new, permanently heightened default state.

Persistent Attention by Experts

I don't watch sports much. Take basketball, for example. I have only the most basic understanding of the rules. I am unable to appreciate much of the action on the court beyond noting whether a ball makes it into the basket or not. However, if I listen to the commentator—who most likely is an ex-player—I am astounded at everything he describes. It is as though he is seeing a totally different scene. He describes and analyzes details involving offensive tactics, defensive zones, clock management, foul strategies, and many, many more nuances I don't see because *I do not know what to pay attention to.* Because the commentator *does* know what to pay attention to, he *is* in fact seeing a different scene. On the other hand, I play some music, which tunes me into the sounds of the instruments I play. This lets me appreciate nuances a performer brings to his craft. Like the basketball commentator, I have learned what to pay attention to.

An auditory expert might be a musician, a bilingual, an athlete, a sound engineer or designer, even a bird-watcher or a meditator. We are all experts of the language we speak. For each variety of auditory expert, the signals on the outside (sound) mold the signals on the inside (electricity). The principles at work in expert listeners apply to all of us. They are just easier to spot in experts, which is why experts can tell us a lot about the brain. How our brain processes sound right now as you are reading this book, an hour from now as you are walking the dog, a week from now as you are cramped in coach on your way to your cousin's wedding, is a product of the sounds we have paid attention to in the past. A lifetime of auditory learning of any kind *cumulatively* shapes our brains. An accumulation of sound experience changes our brains beyond furnishing in-the-moment attentional shifts that aid in accomplishing one-off tasks. The more we pay attention to something explicitly, and the longer we spend on it, the more the sound-coding systems in the sound mind change accordingly.

We Learn What We Care About

We probably don't remember Mrs. Buthan's fifth-period English class with endless sentence diagramming drills because it was boring.* In most cases, we learn what we care about. There is absolutely nothing more motivating, when we are trying to learn something, than feeling strongly about it. Whether you are an owl learning how to find prey or a teenager picking up an electric guitar for the first time, you are activating reward centers in the brain as you attach meaning to those particular sounds. The owl cares about his hunting prowess—his very existence depends on it—and the budding musician is emotionally invested in making her music.

The limbic system dramatically facilitates learning, inducing faster and longer-lasting effects.[41] In fact, reorganization of the sound mind may not take place without it.[42] Reorganization of tonotopic maps in the brain can be achieved by stimulating the limbic system directly with electrical current, even without training. Just pairing a tone with limbic system stimulation changes the auditory cortex's frequency map to one with an outsized representation of that tone.[43] Much as stimulation of limbic areas promotes changes in the auditory pathway, sound alone can excite the limbic system when that sound signifies an event the animal cares about.[44] There is a decidedly two-way street between feeling and the sound mind.

From Conscious to Unconscious Processing of the Sounds Around Us

The other day I changed the ringtone on my cell phone. At first when my phone rang, I didn't hear it right away. After a few days, though, I knew my phone was ringing even when it was in another room.

---

*Truth be told, I loved diagramming sentences.

This is a trivial example of unconscious learning. A much more dramatic example is the famous case of HM, a young man who was suffering from seizures. To relieve the seizures, he had brain surgery that, among other things, removed his hippocampus, a chief site of memory. Although his seizures were relieved, he could no longer form new memories. He would forget people and events as soon as they happened. Yet if he was asked to perform a task like mirror-image drawing, even though the next day he had no recollection of having performed the task, he nevertheless got better at it day after day.[45] He learned *unconsciously*.

Once upon a time, we exerted deliberate and focused concentration to control slippery pedals and wobbly handlebars. Now we automatically, unconsciously, and effortlessly ride our bikes. Sounds that are important to us make that same transformation. Whether in-the-moment or over an extensive period of time, the hearing brain tunes the sound mind. First, the auditory cortex, the most malleable auditory structure, changes so the immediate task at hand can be accomplished. But with continued attention and repetition, structures all along the auditory pathway eventually change so that they achieve a *new default state*. Now, signals that have become important—the sound of our own musical instrument, the sounds of our language, our coach calling out plays from the sidelines, a basketball dribbled up the court, the sound of our name or our new ringtone—are preferentially encoded. Your experience with sound has left a legacy on the sound mind. You no longer have to devote attention to that hard-earned sound-to-meaning connection you once learned to make; your brain now automatically and unconsciously processes sound in a new, efficient, and more rapid manner. Our hearing brain implicitly picks up on sound patterns throughout our lives, beginning in utero.[46]

The more you do something, the stronger the learning in the sound mind. The changes observed in ferrets after a few hours of learning have a less permanent impact on the hearing brain than a lifetime of making music or speaking a second language.

How does our implicit and explicit experience with sound get transformed into memory? The efferent system makes learning possible by changing how sounds are processed by the brain. But not all structures are affected in the same way. In general, the more peripheral (nearer the ear; closer to the bottom of our block diagram) the structure, the longer it takes to change, and the greater the amount of training, practice, and attention it will take. After learning has been accomplished, cortical map expansion can revert to a pre-training state. This transformation happens as learning takes on new strategies that no longer require the cortical effort.[47] But, in more peripheral, subcortical structures, the result of the training—the shift to a new default state (the *memory*)—tends to be longer-lasting.

Therefore, while cortical reorganization contributes to *short-term* memory, *long-term* auditory memory requires a system-wide resetting of the default state of the entire integrated sound mind. This reset involves responses to sound all along our ear-to-brain pathway. That is, the learning-modified activity in the afferent pathway now constitutes memory itself. In this view, each part of our auditory brain houses memories of our sonic experience.

We are not generally aware of the miracles that unfold inside our brains. However, biological principles position us to better understand how the brain forges our own unique responses to sound. Just as the basketball announcer and I see different scenes on the court, no two people experience an auditory scene the same way. Each of us, through experience with and attention to sound—the languages we speak, the music we make, the sounds that are important to our lives—has forged a unique and automatic sound processing infrastructure.[48]

Sound changes us.

Figure 3.5
The sounds of our lives shape our sound minds.

4

The Listening Brain: A Quest

Science is a deeply human endeavor.

"What is going on in the brain?" comes up again and again. It is at the heart of everything I study. The biological underpinning of sound in language, music, and health is difficult to gauge if you can't measure what's going on in the brain. I have spent years searching for a satisfactory view into the subtleties of sound processing our sound minds are capable of.

Scientists stand on a narrow plank and, over decades or generations, build outward trying to extend the bit of floor we can reasonably trust to hold our weight. A given plank can look promising for a while, but if it begins to get rickety, it may be abandoned for another pathway if it fits the known facts better. Science is a deeply human endeavor. It is a humble attempt to cast a little light into the vast darkness of our ignorance. My journey—the plank I have aimed to contribute—is finding a window into sound processing in the brain.

Scientific progress is not a collection of facts. It relies on *context* and on *people*. Science often comes to us, the public, in the form of uncontextualized sound bites or headlines: "Science reveals that bacon is good for us." Last year's headline that bacon is bad for us is forgotten or rendered inoperative. This is not how "real" science proceeds. Science works by the slow accretion of ideas that have withstood repeated testing. Those two bacon studies, together with scores of studies that came before and all those that are to come, add to a growing body of evidence that cumulatively sheds light on the health and nutritional value of salt-cured pork belly. No one study in isolation should tempt us to declare "case closed." Journalists and, on occasion, an individual scientist hoping to obtain funding or notoriety, have an incentive to present the latest findings as though they disposed of the matter. But as satisfying as it might be to be able to "wrap things up," it is harmful when carelessly reported scientific findings lead people to conclusions that suit them while ignoring those that do not.

Historically, hearing science has focused on the ear-to-brain direction. It made sense to start at the beginning (the ear) and build on that understanding, gradually putting planks together to figure out how sound makes its way toward and through the brain. As the field evolved—and Brainvolts has been part of this shifting dynamic—we have come to realize the ear-to-brain system is only part of a much deeper system that engages much of the rest of the brain.

I am always desperate to know *what is going on in the brain* to figure out what goes on in the sound mind based on our lives in sound. My quest is to learn how we might sculpt our sound minds to make us better musicians and athletes, to hear all kinds of things better, from the song of a bird to the whispers of our loved ones.

1. What I needed was a biological approach that **could reveal sound processing so subtle we aren't even conscious of it.** An experiment performed in the hippocampus (important for forming new memories) gave me inspiration. Fried and colleagues made direct recordings from the hippocampus while people were shown a set of pictures. They discovered that hippocampus neurons responded when the participant looked at

pictures he had seen before *but didn't remember seeing.*[1] The brain clearly "knew" more than the person was consciously aware of. I was looking for the sound-mind analog of that.

2. I needed to **capture how the sound mind processes sound ingredients like pitch, timing, timbre . . .**

3. I needed **to get this information without requiring active participation from the listener.** A sound-mind probe needs to work in people who might have difficulty performing a task, are too young or sick to sit quietly, or those for whom there is a language barrier. A sound-mind probe that defies gaming the system. I wanted a unified approach that works for everyone.

4. I needed a probe that reflects how the sound mind is **shaped by experience** such as learning another language, making music, being an athlete, struggling to read, or sustaining a brain injury.

5. Most of all, I needed a probe to reveal sound processing in an **individual brain**, one that shows how someone hears the world in their own unique way.

Today, I know the frequency following response (FFR), a method that captures the brain's response to sound, can provide all of these insights into our sound minds. What follows is my journey to developing this approach to serve our questions, with false starts and dead ends along the way—the story of turning what we did not know into what we now know, and the questions we are now in a position to ask.

Measuring the Signals Inside the Head from the Outside

If I were talking to you now, the neurons in your auditory brain would be producing electricity. The electrical responses to sound that make their way to the surface of the scalp are tiny, but we *can* measure them with scalp electrodes. To do so is challenging. The brain is generating electricity in response not just to the sound but also to what we are looking at, the mechanics of sitting in an upright position, our heartbeat, etc. Plus, there are the electrical fields generated by the computer across the room, the electrical sockets in the wall, our smartphone . . . We must pull the tiny electrical response to sound out of the much larger—but for our purposes irrelevant—electrical cacophony existing inside and outside the head.

So what if we could make all the other nonauditory electrical noise go away? We can, at least to a first approximation, by using signal averaging. The idea behind averaging is that the response to a given sound happens at the same time, every time, over many repetitions of that sound. Electrical noise, whether coming from the person or from external sources, occurs randomly in time and will gradually fade away with averaging. The computer is whirring continuously and the person scratches his nose whenever it itches and his heart is beating away. But the sounds keep playing all the while. The computer playing the sounds pinpoints exactly when the sound was played, and the responses are all stacked together with respect to the sound onsets. Thus, with this approach, any brain activity that is synchronized to the sounds of interest *constructively* contributes to the final average. At the same time, unsynchronized noise events—the cough, the knuckle crack, the flickering fluorescent light—*destructively* mingle until, with enough repetitions, they average out and approach zero. Once the noise interference is small enough, we are left with a snapshot of what the sound "made" the brain do.

Without the electrodes ever moving from the surface of the scalp, it is possible to change the sound and pick up activity from different points along the auditory pathway, from the auditory nerve to the cortex. But wait. If we do not have an electrode directly in one of the auditory structures, how do we know whether our recording comes from the brainstem, midbrain, thalamus, cortex, or elsewhere? Inferences can be made based on principles of the auditory pathway discovered by recording directly from those regions; mostly, it boils down to speed. The speed at which neurons are able to synchronize (fire together to sound ingredients) decreases as the ear-to-brain ladder is climbed. Some auditory structures are specialized for dealing with timings in the tens of seconds, others for seconds or milliseconds or microseconds. In short, the cortex is slow and the subcortex is fast.*

Knowing When Sound Changes: Step 1

Scientists have capitalized on the fact that the brain will respond to a *change* in an otherwise predictable pattern, whether auditory, visual, or somatosensory. To test for the detection of a change in sound, you play a repetitive sound and every now and then, say 10 percent of the time, you substitute a different sound. Beep-beep-beep-beep-beep-beep-*boop*-beep-beep-beep. After the "boop" you get a perturbation in the electrical waveform recorded from the scalp signaling the brain has detected the change from beep to boop. This important and practical life skill likely evolved as our most distant ancestors needed to detect changes in the ongoing soundscape to be alerted to potential sources of danger (the sudden movement of a snake while crickets are chirping). Detecting changes in sound is thus deeply ingrained and worthy of study.

---

*The auditory nerve can keep up with high speeds of firing. If you play a tone at several thousand hertz (that is, several thousand sinewave cycles per second), the auditory nerve will fire robustly to each cycle. On a good day, the thalamus taps out at a few hundred hertz and the auditory cortex at about a hundred, with the midbrain somewhere in between. Consequently, if you record a robust 700 Hz response from a scalp electrode to a speech sound (700 Hz is an important harmonic frequency of a typical "ah" sound), you can confidently rule out a thalamic or cortical origin. The thalamus and the cortex still process these higher frequencies, but they do not do so with the rate-locked coding characteristic of, for example, the midbrain, which phaselocks (fires) at every cycle.

The most widely known response of this type has even been used as a way to extract information in criminal investigations. It goes like this: let's say a murder has been committed. Suspects are wired up to electrodes and shown pictures of various weapons one after another—a handgun, a rifle, a tire iron, a bottle of strychnine, a hunting knife, a meat cleaver, a hammer, etc. The brain of an innocent suspect, who has no knowledge of the crime, would have the same physiological reaction to each picture. But the guilty party's brain would fire off a distinct response to the weapon he used to commit the crime.[2]

In the late 1980s I attended a conference in Hungary, and a Finnish neuroscientist named Risto Näätänen got my attention. He also inspired me by swimming in the icy lakes in Lapland, way in the north of Finland, while I stood by in my down parka and hat. He emerged booming with energy twenty minutes later. I eventually found my way into one of those lakes, but only for a few seconds, and only with a well-formed plan to retreat to a sauna nearby.

Risto discovered he could see the brain's response to a change in a sound pattern *even if we're not paying attention to it*. He called this response the "mismatch negativity" or MMN.[3] It is a downward-going brain wave (hence negativity) that occurs when a sound doesn't match the rest of the sounds in a sequence (hence mismatch).[4] What was remarkable was the response occurred automatically; that is, it did not require explicit participation from the person hearing the sounds—hitherto a requirement in change-detection research. Instead, his research participants could be reading, watching a captioned video, sleeping, daydreaming, or otherwise ignoring the sounds. Here was a response that met one of my requirements; it was passively obtained. **No need for active participation from the listener**.

The sound changes were readily detectable if you paid attention to them. But I was sitting there wondering, what if we could go a step further and measure the brain's response to a change in sound that *wasn't* detectible? A change so small we could barely detect it even if we tried? We already knew that children with language problems had difficulty processing sound. I suspected they might have difficulty processing nuances, such as the fine-grained differences between speech sounds. How could you get a toddler to tell you what sounds they can and cannot distinguish? It can be hard enough to get a young child to indicate what they can hear even if the sound differences are obvious, let alone if there is subtle millisecond timing involved like there is in the sounds of language. What if we could see, biologically, without a child having to respond directly, what distinctions they could or could not hear?

Mikko Sams, another Finnish neuroscientist (maybe those extralong dark winter nights heighten Finns' consciousness of and interest in sound), looked at brain responses to subtle sound changes—a 1,002 Hz tone compared to a 1,000 Hz tone.[5] The MMN proved that the brain was capable of discerning that 0.2 percent difference. Still, with effort and concentration, someone can perceive that tiny difference. So next at Brainvolts we made the difference even more challenging. Would the auditory brain respond to *physical differences in sound so small a person cannot consciously detect them even if they try*? This time, we created syllable pairs with such subtle acoustic differences our research participants *couldn't* tell them apart. Even though they could not consciously tell them apart, similar to the hippocampus "remembering" the pictures that had been seen, their sound minds still could![6] We now had a brain response that satisfied our second condition, that of **reflecting perceptions so subtle we are not consciously aware of them**.

Using the mismatch negativity, we discovered the brains of children with language disorders could not distinguish the minimally contrasting speech sounds a typical child's brain could distinguish. This uncovered a biological bottleneck facing these children. Language difficulties, we surmised, could stem from a failure to connect the subtle sounds of language with their meanings. If this were true, language development could be guided by strengthening processing in the sound mind.

The limit of what our brains can and cannot distinguish is not set in stone. Like any other system, we can push our limits through training. What if we start out not being able to register a difference in a given pair of sounds, but then train ourselves to distinguish them? Would the MMN emerge or grow with learning? Brainvolts graduate student Kelly Tremblay tested this question by teaching people to hear sounds that do not occur in their native language. These were sounds an English-speaking listener did not initially distinguish, yet were readily distinguishable by speakers of other languages. Sure enough: with training, the brains of English speakers began to show signs of making the distinction between the sounds well before they could consciously tell the sounds apart.[7]

This got me thinking about the potential for using a similar approach with children who have language disorders. And I was inspired by the possibility of objectively monitoring a child's progress, using a probe of the sound mind to gauge whether appropriate rewiring was taking place even if it wasn't yet evident in the child's behavior. As I play the piano every morning, I like to think my brain is learning even if the notes don't sound any better than they did yesterday. I am encouraged that eventually my fingers will catch up with my brain.

Although it advanced my thinking about sound processing in the brain, the mismatch negativity approach was ultimately unsatisfying. First, the electrical activity we are interested in is easily dwarfed by the electrical activity associated with blinking, or muscle tension, or throat clearing. When I was looking at a negative-going waveform, I sometimes couldn't convince myself I was looking at a response to sound rather than a sniffle, because the slow waves of the MMN blended in so well with other electrical signals. Brainvolts even published a paper called "Is It Really a Mismatch Negativity?" devoted to strategies for wringing this response out of the background noise.[8] Second, working with mismatch negativity is slow-going. It is predicated on something happening a small percentage of the time. If only one of every ten sounds is going to elicit the response, recording it is inescapably going to take a long time. This is impractical and a problem when working with children and in the clinical settings I envisioned. Third, because the MMN is mostly a cortical response consisting of slow brain activity, it doesn't reflect the many fast ingredients inherent in sound. All we had was a neural deflection that signaled the brain had detected a changing soundscape. It did not tell us how the brain responds to the many ingredients, slow and fast, that make up most sounds. It was time to move on.

Processing Sound Ingredients: Step 2

Around the turn of the century, I gently began a course correction for Brainvolts that would become pivotal to our work. All along we had been measuring responses to sound with scalp electrodes in humans while in a parallel line of research, we were measuring activity in the auditory structures of the guinea pig brain. We continued to use methodology similar to what I had used in my first learning experiments in rabbits. It was now time to connect the past with the present.

Doctoral students Jenna Cunningham, Cindy King, Brad Wible, and Dan Abrams made recordings in both cortical and subcortical structures of the guinea pig brain. Using speech sounds, they were picking up beautiful, clear responses within the midbrain, thalamus, and auditory cortex, including the fast and slow activity patterns visible in each of these structures. But we noticed something else. Simultaneous to these deep-brain recordings, we always kept an electrode on the surface of the brain as a way to connect what we learned from activity inside the brain to what could be measured from outside. From that electrode, which was not very different from the type we use in humans, we were able to see a pretty clear representation of the acoustic ingredients present in a complex sound wave! Like the responses we were pulling out of the midbrain and thalamus, this surface-recorded brain wave was rich enough that we could analyze it and determine whether the sound was a "ba" or a "pa." We could look at the brain wave and deduce "a" or "oo." The approach was quick and practical. A single brain response to a single speech syllable from a scalp electrode revealed a wealth of independent pieces of biological sound processing, in part because all of the ingredients in the sounds that are important to us—pitches and FM sweeps and harmonics—were present and accounted for.

This led to discussions with the team, including my longtime collaborator Therese McGee. We agreed this type of recording process had the promise to meet another of my requirements, of being able to **capture how the sound mind processes the ingredients that make up sound** using the brain's rich anatomical and physiological infrastructure. It was an approach I could sink my teeth into. It was an approach that brought me back to the signals I have always gravitated toward for reassurance and insight.

This brain activity is called the frequency following response, or FFR. While not particularly new (FFR was discovered in the 1960s),[9] its ability to do more than signify the detection of sound—usually just a single tone at that—was not appreciated until later. But even in the 1990s, when more complex sounds began to be used,[10] the FFR was only used to probe how the brain processes the fundamental frequency of sound—just one of many sound ingredients, fast and slow, that make up our sonic world. Brainvolts ran with the idea that the FFR can also reflect the brain's processing of the many rich sound details we need to make sense of sound. In fact, the brain response is so precise that it physically resembles the sound wave that elicits it. You can see the detailed sound ingredients in the brain response itself (figure 4.1).

Most brain responses to sound don't tell us much about how the brain processes sound *ingredients*. They are analogous to lipid panels. High cholesterol is statistically predictive of atherosclerosis. But your cholesterol level is not an actual measurement of the narrowness of

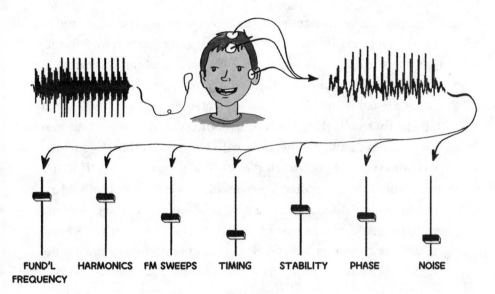

Figure 4.1
Sound is delivered to the ear through earbuds. Scalp electrodes pick up the brain's electrical response to sound. The brain wave resembles the sound wave. It tells us how well the brain processes different ingredients of sound. The mixing board illustrates that each sound ingredient is processed distinctly.

your artery walls. fMRI (functional magnetic resonance imaging) and many neurophysiological responses, like cholesterol measurements, enable us to make inferences about something that might be going on in the body, but they do not offer the precision that a direct measurement of arterial plaque would provide. Most physiological responses to sound do not show us how pitches and timbres and loudness and movements in time all work together; they're just . . . abstract bumps and spikes. Imagine the possibilities of a brain response that actually resembles the sound itself. Now imagine being able to directly measure how each of us processes these *ingredients* in the way that makes us unique. This is what the FFR can do. With it, we can peel away levels of abstraction endemic to biological testing. I cannot think of another biological response that is so close to a one-to-one representation of processing in the brain—it's practically unheard of.

To investigate how our sound mind makes sense of sound, I explicitly wanted to use interesting sounds like speech and music, applause and dog barks and crying babies . . . and now (as opposed to using FFR to capture just the fundamental frequency), all of those sounds are fair game. The response to them is a well-defined signal that clearly recalls the speech or bark or cry that evoked it and, in so doing, represents the precision with which the brain encodes it.

We can see how good a job the brain does processing each sound ingredient. These ingredients are not all processed identically like a volume knob. Rather, think of a mixing board, where the processing of each ingredient tells part of the story. The faders on the sound-ingredients mixing board reveal specific enhancements and bottlenecks that characterize certain groups of people and individuals based on what they were born with and how they live their lives in sound.

Listening to the Listening Brain—Art and Science

Because the brain wave resembles the sound wave used to evoke it, we can actually *play back* the brain's response to sound and *listen to the brain* (figure 4.2). Brainvolts has recorded brain responses to many sounds, including several octaves of notes. It is possible to

Figure 4.2
A microphone turns a sound wave into an electrical signal that can be played through a speaker. Likewise, listening to sound causes neurons to fire, creating electricity in the brain, which in turn can be played through a speaker. Sure enough, when sonified, the frequency following response sounds pretty much like the sound that evoked it, albeit a bit muffled.

transfer the brain's response to each note onto a "brain keyboard." As we "play the brain," it is easy to hear how each person processes the same notes in their own unique way. If you're curious, you can listen to the examples of brain wave sonification on the Brainvolts website. Occasionally, when I get to share the stage with a musician, it is wonderful to hear the "brain keyboard" interpreted by a piano virtuoso.

Another example of the interplay between art and science took place when I had the pleasure of sharing a stage with opera singer Renée Fleming. While Renée sang her especially moving rendition of Dvorak's "Song to the Moon" from *Rusalka*, I sat in a reverie, listening, on a piano bench near her. When she finished, I had to compose myself enough to stand up and find my way to center stage, I could not speak for several moments—a testament to the formidable power of sound. Fittingly, it was my job that night to explain what happens in the brain when music moves us.

Figure 4.3
Science and art, circa 1997. My son's middle-school rendition of "what mommy does."

I strive to celebrate the art in science. I like to use illustrations when I teach and deliver lectures to help get scientific ideas across, reinforce the beauty inherent in science, and give us a sense of something larger than ourselves.

Experience

While we were figuring out how to pull sound ingredients from the FFR to investigate sound processing in language-impaired children, a report by Ravi Krishnan smacked me over the head. He had found that Mandarin speakers' brains (revealed by the FFR) were uncommonly good at tracking the pitch of sound, something English speakers' brains did not do as well.[11] The Mandarin speakers' sound minds had nudged up the pitch-tracking fader to accommodate the tonal aspect of their language that the English language lacks. This precise, language-specific processing was so ingrained, the Mandarin speakers' brains did it *in their sleep*.

It was clear the speakers of Mandarin had honed their pitch prowess from a lifetime of making sound-to-meaning connections in their native language. Importantly, these experiments revealed a mechanism of action—*how* sound processing was altered by experience. Ravi did not report a vague "lighting up" of this or that brain region. He was not looking at blood oxygenation levels or broad negative deflections in a waveform or a blunt response to a sound's onset. Instead, he described the brain's coding of one isolated ingredient in sound: a pitch tracking difference between two groups of listeners was illustrated unequivocally. In other words, the FFR clearly reflected what was going on in the sound mind; the sound ingredient was *right there* in the neural response.

While Mandarin speakers excel in pitch tracking over the duration of a syllable (on the order of 200 ms—a long time in speech), children with language disorders have difficulty processing rapid cues such as consonants transitioning into vowels (FM sweeps that last only a fraction of that length). Would the FFR be powerful enough to probe that kind of sound ingredient? All the ingredients in speech? Because of its subcortical roots, the answer is yes. The FFR is not subject to the speed limits that hamper the utility of the cortex-centric MMN or the slower process of fMRI.

To move beyond pitch processing, Brainvolts had to get busy. Using the FFR as a measure of pitch (the fundamental frequency) had been around for a while, but no one had thought to look at ingredients like FM sweeps and harmonics. Luckily, there is a reasonably direct path between analyzing sound itself and analyzing a physiological brain response to it—especially, as we have seen, when there is such an obvious similarity between them. The techniques for extracting FM sweeps and harmonics and timing and quantifying noise levels were already well understood in the signal processing world. We just had to apply them to physiology. It took my team learning these techniques and applying them to this new sort of signal to unlock the power of the FFR—adding faders to the mixing board. Over the years we have refined these processes and have published tutorials[12] aimed at delineating the process. It is now possible to compare the signals inside the head (brain waves) to those outside the head (sound waves). That we can measure a brain wave that is close enough to a sound wave even to begin seeing those parallels resonated strongly with me. This precision compared favorably with the precision of my early microelectrodes in rabbits or Cindy, Jenna, Brad, and Dan's guinea pigs. A precise finger on the precise pulse of auditory processing, rooted in sound and signals, had become possible in humans.

The FFR reflects **what our hearing brain has become from our experience in sound**. Brainvolts pioneered looking at the impact of experience and disorder on the processing of sound ingredients. This enables us to understand how our life in sound alters our default physiological response to specific ingredients in our sonic world. The mixing board metaphor gives us a way to understand the strengths and weaknesses that different populations may have and the effects that life experiences may have on sound processing.

Crucially, each person's response to sound is unique. The subtle **differences among individuals** can now be measured, seen, and even heard. A person's sonic history can be told through their response to sound—their biological fingerprint.

Snapshot and Hub of Auditory Processing

In the view of the auditory system as a hierarchical ear-to-brain conveyance, it was hard to imagine the midbrain—from which the FFR largely arises—as a *hub* of a rich, distributed, bidirectional system. In that view, the midbrain is just a way station of auditory processing on its way from the ear to the brain.

The conceptual advance Brainvolts helped to guide was to think of the midbrain as a hub, not as a mere link in a chain of ear-to-brain processing (figure 4.4). The auditory pathway is a loop, and the subcortical auditory centers are not just hardwired conduits for sound. The auditory midbrain is a hub of our cognitive, sensory, motor, and reward networks—at the heart of this constantly evolving, distributed neural infrastructure of sound processing.

The idea that the midbrain could reveal sophisticated aspects of sound processing was overlooked in part by the pervasiveness of brain imaging such as fMRI. Imaging excels at revealing cortical activity (and does so in a visually satisfying way), thereby fueling the view that to understand how the brain makes sense of sound, one must focus on the cortex. The FFR, probing the sound processing in the subcortex with exquisite precision, provides a snapshot of the activity of the whole sound processing network, from stem to stern, that is our sound mind. Said another way, if your back hurts, the pain could arise from a problem with your knee. Similarly, although a predominant

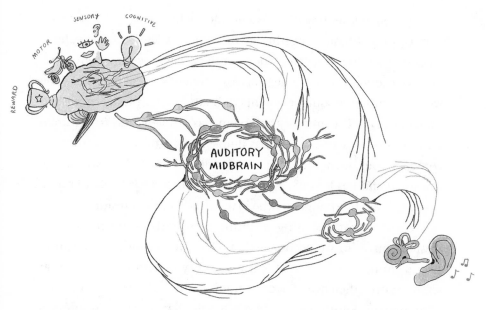

Figure 4.4

Influences exerted by learning create an auditory system that is flexible. Rapid changes in sound processing to satisfy an in-the-moment demand originate in efferent (dark) pathways and eventually effect permanent change to afferent auditory pathways (light), resulting in a new default state. This is how sound memories are stored.

source of the FFR is the midbrain, it should not be interpreted as a "midbrain response." *The midbrain is in the middle of the action.*

There is an ongoing conversation in neuroscience and philosophy called the *binding problem.* It boils down to a question of how the brain coordinates all its inputs, the sights, sounds, smells, tastes, and touches, guided by a lifetime of accumulated experience, into a concrete whole.[13] How does the combination of ever-accumulating sensory input produce the knowledge: "That's my phone ringing" or "I hear my brother pulling into the driveway." Where does the necessary unity come from? Somehow, the brain gathers information and "binds" it into a unified perception.

V. S. Ramachandran describes experiments that "flatly contradict the theory that the brain consists of a number of autonomous modules acting like a bucket brigade." As Iain McGilchrist puts it, "Experience is not just a stitching together at the topmost level. . . . Perceptions emerge as a result of reverberations of signals between different levels of the sensory hierarchy, indeed across different senses."[14] Much of the work of uniting modular elements of brain function is accomplished subcortically.[15] The auditory midbrain has ample access to information from other senses, as well as limbic and cognitive input from all corners of our distributed and interconnected brain. This knowledge is learned and has become automatic. Thus the FFR is likely to reveal how the brain binds together the many dimensions of hearing.

We know that assigning meaning to sounds—learning—brings about changes in sound processing. First, we find out that a sound has meaning, then we sculpt the auditory system to process that sound more efficiently. Specific brain centers, auditory and otherwise, operate jointly to contribute to the default response properties of the midbrain. The FFR is thus *far* from a reflection of activity from any single auditory structure.[16] Remember, the hearing brain is vast. Specific brain centers, within and outside the auditory pathway, each make their own contribution, but those centers function jointly, and in the context of broader neural networks. The FFR gives us a functional view of sound processing in the brain. It provides a snapshot of *how well the entire sound mind is coding sound ingredients.*

The search for a viable biological portal into sound processing has played a role in the evolution of my thinking about the hearing brain. It has helped me see sound processing outside compartmentalized, assembly-line brain centers. It has provided a way to embrace the vastness of the hearing brain with its sensory, cognitive, motor, and reward networks and to think about our lives in sound more holistically.[17] Perhaps this backstage tour of one science lab's quest can provide a glimpse of how scientists try to build a plank solid enough to form a firm floor. Our experience positions us to consolidate what we now know, articulate what we still don't know, and home in on what we are striving to understand about the sound mind.

II   Our Sonic Selves

5

Music Is the Jackpot: Sensing, Thinking, Moving, Feeling

If it feels good, it sounds good.

—Salvatore Spina

The Musician Brain

A doctor who was present at Beethoven's autopsy noted that his "brain convolutions appeared twice as numerous and the fissures twice as deep as in ordinary brains." Schumann did not fare as well; his doctor noted "considerable atrophy of the whole brain."[1]

In the early 1900s, a more systematic investigation into the structure of the musician brain was undertaken by German surgeon Sigmund Auerbach. Unlike some of the claims during the early twentieth-century transitional era in medicine, which included the massive consumption of grapes as a cure for cancer[2] and transplanting goat testicles into men as a remedy for impotence, Auerbach's work was rooted in the scientific method. He reported the brains of noted musicians, examined after death, were larger in temporal lobe regions, including a part of the auditory cortex, than those of nonmusicians.[3] Auerbach, who went on to make contributions to the treatment of epilepsy and brain tumors, concluded those brain regions accounted for the musical skills of those musicians. This launched a host of investigations that bore out that the musician brain was indeed structurally distinct from the nonmusician brain. Structural differences have been noted in the auditory cortex,[4] the somatosensory cortex,[5] the motor cortex,[6] the corpus callosum,[7] the cerebellum,[8] and the white matter tracts within the cortex[9] and connecting subcortical and cortical brain regions.[10]

We do not know that Beethoven's remarkable convolutions, or any findings about brain structure have bearing on how musicians' brains *work*. Functional rather than structural differences are what matter most. Musicians have stronger cortical responses to musical instrument sounds than nonmusicians.[11] Musician brains more readily register a change in a sound pattern or a dissonant or mistuned chord.[12] The brains of rock guitarists respond powerfully to power chords.[13] Certain sound *ingredients* are strengthened in musicians, notably harmonics, timing, and FM sweeps as we'll get to in detail later.[14]

Music Engages the Sensing, Moving, Feeling, and Thinking Brain

The sound mind is vast, engaging our cognitive, motor, reward, and sensory networks.[15] Music does an exceptional job engaging these systems, providing effective avenues for learning through sound (figure 5.1).

Sensory: *Auditory*

Making music alters the sound mind's default, automatic response to sound—our fundamental auditory selves. Playing music positions us to build a brain especially tuned to our sonic world.

Mari Tervaniemi was one of the first to show differences in neural sound processing between musicians and nonmusicians and

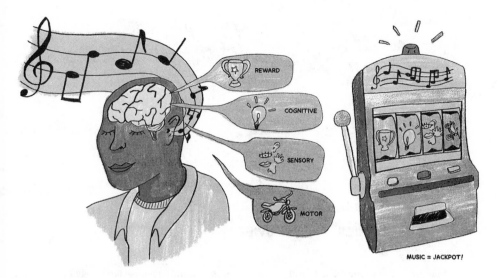

Figure 5.1
Music is the jackpot of sensory, cognitive, motor, and reward engagement through sound.

between different types of musicians.[16] If you play a five-note melody to a listener several times, deedle deedle dee, deedle deedle dee, deedle deedle dee, and then suddenly change the melody, deedle *doo*dle dee, the brain notices and signals this change with the MMN (mismatch negativity) even when the listener is ignoring the melodies. Mari found that musicians' responses to the new melody were enhanced relative to the nonmusicians.[17] She went on to demonstrate this musician-brain enhancement to pitch, timbre, duration, intensity, roughness, location, and the rules of harmony.[18]

Harmonics, timing, and FM sweeps are at the heart of the "musician signature" response to sound (see figure 5.2). Making music strengthens the sound mind across the lifespan and builds up over the years.[19] Importantly, it changes how the brain responds to sound in general, not just to music, notably to speech.

Now is a good time to address two questions I am often asked about the musician brain. The first is, "How do you define a musician?" For the purpose of defining how much music-making it takes to influence

HARMONICS   FM SWEEPS   TIMING

Figure 5.2
Making music strengthens sound processing in the brain. The musician's edge builds over a lifetime.

the sound mind, the answer is simply someone who plays music regularly. The musician doesn't need to be especially proficient. "Regularly" can mean making music for as little as a half hour a few times a week.

Next, I'm asked, "Does the instrument I play matter?" The answer is no and yes. "No" because the brain signature of heightened processing of timing, harmonics, and FM sweeps exists regardless of the instrument you play, including your voice. "Yes" because the sounds of the instrument you play will be *especially* well processed by the sound mind. Brain imaging of violinists and trumpeters reveals that each group preferentially codes the sound of their own instrument in the auditory cortex.[20] This finding was replicated, pitting violinists against flautists,[21] and extended into midbrain processing of sound ingredients as shown in figure 5.3.[22] That is, the sound of a piano is enhanced in pianists; the sound of a bassoon is enhanced in bassoonists, and so on. Moreover, conductors have an exceptional ability to localize sound arriving from all corners of a room.[23]

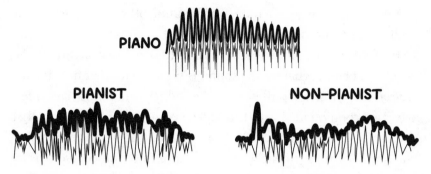

Figure 5.3
The musician's auditory brain responds exceptionally well to the sound of their own instrument.

Sensory: *Auditory-visual*

When making music, seeing is tightly bound to listening, from watching your band for cues, to following the directions of a conductor, to reading music. Playing music can enhance visual processing and, especially, combined auditory-visual processing.

Collegiate drum corps consist of percussionists, brass instrumentalists, and the color guard. The latter do not play instruments but instead perform intricate movements with flags, rifles, batons, and sabers to provide a synchronized visual display complementing the accompanying musical performance. They are well practiced in throwing and catching a whirling flag after a precise amount of time has elapsed and a specific number of rotations has been completed. And they do so in sync with dozens of their cohorts. One would expect the color guard to perform especially well on tests of visual skills, but this is not the case. The visual skills of championship-caliber color guards were not as good as those of the two musician groups, especially the percussionists.[24] Thus, it seems likely that the degree to which visual timing is honed by years of music making outweighs direct experience with activities requiring visual timing by itself.

When someone hears a note of an instrument, such as a cello, their auditory brain produces an electrical signal that resembles the sound of the cello, visible in the FFR. A musician's response is slightly faster, richer, and larger whether the sound is heard by itself or heard while *watching* a person play the cello. The musician/nonmusician differences are compounded with the added visual input,[25] suggesting that the interconnections between the auditory and visual systems in the course of engagement with music have finely tuned audiovisual skills. This discovery was Brainvolts' first publication on the effects of musical experience on the sound mind. While we were not surprised musicians would have enhanced audiovisual responses to music, the unexpected discovery was that this audiovisual enhancement was also seen to *speech*. Similar to the cello phenomenon, this was evident when a heard voice was accompanied by watching a person speaking.

Gabriella Musacchia, a trumpet player, teamed up with our guitar-playing Finnish collaborator Mikko Sams for this work. Gabriella now has her own laboratory, having founded a drumming program for toddlers in New York City along the way.

Moving: *Auditory-motor*

"Pay attention to your fingering!" says my piano teacher (again). "When you move your hands easily and with precision, the music *sounds* better."

Robert Zatorre is one of the most prolific and influential scientists to investigate the effects of musical experience on the nervous system. His group discovered that the motor cortex is active when we *listen to music without moving*.[26] And in musicians, even just *thinking* about playing music activates the motor system.[27] This illustrates the tight link between our hearing and motor systems, especially in people who play music.

Right-handed people, by virtue of writing, brushing teeth, and performing other tasks with their right hands in the course of their daily lives, come to have asymmetrical cortical motor maps.[28] Specifically, the left motor cortex—which controls the right hand—is simply more developed. The converse is true for left handers. However, professional keyboard players, who have developed well-honed, precise skills in both hands, have symmetrical motor brain maps—driven by an *expansion* in the map that controls the nondominant hand.[29]

Unlike keyboardists, violinists and other string players engage their motor systems in a decidedly *asymmetrical* fashion. Compared to her right hand, a violinist must have a very dexterous (ha!) left hand. She must make fast and independent finger movements to the correct locations on the correct strings to play the correct notes. The right hand is of course active too, but those movements do not require precise and independent finger movements. Thus, we have an ideal situation for the scientist: the within-subject control. We can examine, in the same violinist, the motor and somatosensory maps that correspond to the left and right fingers. Sure enough, in violinists, cortical regions that control the fingers of the left hand expand, taking brain real estate away from the area that typically is mapped to the palm. No such expansion of finger territory is found for the regions controlling the fingers of the right hand.[30] Additionally, the extent of left-finger expansion is correlated with the length of playing history, which probably rules out a genetically determined extra-large left-finger map that predated the onset of violin playing.

When you're playing music, you get the equivalent of target practice, where the bullseye is creating the sound you want. This involves making a series of comparisons between the sound you're making and the one you're actually after. The practice comes from coordinating your movement with the timing features of your auditory environment, be it a metronome or other people. Sound and movement fuse into a nonverbal form of thinking and knowing. We see it in the brain.

Feeling: *Auditory-reward*
Sometimes when I wake up in the morning, I don't feel especially hopeful. Playing the piano even for just a few minutes, though,

encourages me. By the time I'm getting on my bike to go to work, everything *feels* nicer.

Music has been called the language of emotion.[31] Some of the first connections between parents and their babies are with songs. There is a rich and convincing scientific literature that backs up the music-emotion link. For one, emotional responses are accompanied by certain physiological reactions such as changes in skin conductance (sweat), facial expressions, heart rate, blood pressure, respiration rate, and skin temperature. Music can evoke all these reactions.[32]

Music activates the reward circuitry of the brain. The brain bases for emotional reactions are seated in the limbic system that includes the amygdala, nucleus accumbens, and caudate nucleus.[33] The emotions evoked by listening to pleasurable music activate the very same brain regions that respond to food, sex, money, and addictive drugs.[34] In a study I find particularly compelling, the Zatorre lab found dopamine was released in subdivisions of the limbic system during the *anticipation* of a musical climax as well as *during* the musical climax itself.[35] Not only music but the mere anticipation of music is emotional. I think it is similar to the response we get when we are away and thinking of home. Music builds harmonic tension and spirals sometimes far away, but eventually the harmonic resolution brings us back home. In another examination of emotion and music, people were called upon to listen to new pieces of music and subsequently value them by deciding how much they would pay to hear a given song again. The amount they were willing to pay was retrospectively predictable by the amount of activation observed in the limbic system on that first listen.[36]

Some people actively dislike music, or at best are staunchly neutral about it. This is called musical anhedonia (absence of hedonism). These people have otherwise typical reactions to sex, food, drugs, and money—so, are not depressed or experiencing other conditions that cause broad-based emotional flatness; they simply and selectively do not care for music. This indifference is backed up by a lack of physiological reactions to music such as the change in skin conductance and heart rate that go along with pleasurable emotions.[37] In musical anhedonics, there is reduced activity in the limbic system during music listening; however, they have typical activation levels during gambling for money.[38]

Even before we register the words that are spoken, we react to the voice of people we have emotional attachments with. This is because of the sound to (emotional) meaning connections we have made with them over time. At Brainvolts, we wondered whether the sound minds of musicians might be more sensitive to emotional sounds like a baby crying. We learned that musicians are more attuned to the emotion-bearing harmonic components of the cry while nonmusicians put their neural energy into picking up the voice pitch (fundamental frequency), outprocessing the musicians on this sound ingredient.[39] In contrast, the musicians "conserved" neural energy and responded vigorously only to the most meaningful part of the cry (figure 5.4)—the part that tells you if your baby needs his mommy, or whether it would be better for both of you to let him cry for a while.

Thinking: *Memory and Attention*

I learn so much from my children. My middle son, now in his thirties, is a wonderful pianist. When he was about seven years old, I noticed he was playing the piece he was working on without the sheet music and said, "How wonderful sweetie, you learned that by heart!" Without missing a beat, his reply was, "No, Mommy, by brain." Touché.

Playing music by heart—er, by brain—requires focused attention and memory. We need to *remember* sound patterns, notation, fingering patterns, names of notes, musical terms, and musical expectations (keys, modulations, themes, harmonic relations). Memory enables us to pick up a piece and play it, even from an arbitrary starting point, or to play by heart. *Attention* is also indispensable. Attention is employed

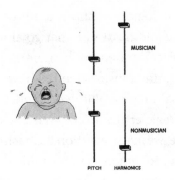

Figure 5.4
Auditory processing of an emotional sound. The nonmusician brain is focused on
the fundamental frequency; the musician brain emphasizes the harmonic content.

to listen to the sounds you are making, to adjust on the fly as needed;
match tempo and dynamics when playing with a group; focus on
the score; block out distracting sounds; focus on fingering, bowing,
embouchure, and breath control; and endure long practice sessions.

Playing music exercises attention and memory. And, like any other
skill, exercise leads to improvement. Thus, it is reasonable to expect
that music playing might be an exercise that would strengthen these
cognitive abilities.

As you are reading this book, your ability to make sense of it relies
on your working memory. You need to remember what you just
read to make sense of what you're reading right now. When you are
speaking to someone, you must "follow the conversation" to make
the interaction worthwhile. Working memory makes these things
possible. Evaluating auditory working memory generally includes
recalling a list of words with some kind of manipulation, like listening to a list of animals and repeating back only the mammals, or
resequencing the presented list in some specific way.

If a musician is trying to learn a musical passage, by seeing that
passage in notation, hearing someone else playing it, or hearing a
recording of it, it's important she be able to hold some model of the
sound she is trying to approximate in her mind as she is resolving

the physical complexities of playing the passage. On the whole, musicians outperform nonmusicians in a wide variety of tasks of verbal memory,[40] working memory,[41] and sequencing (figure 5.5).[42]

On the attention side, musicians also typically perform better than nonmusicians.[43] These tasks usually involve rapidly switching from one task to another or needing to react to a target sound while refraining from reacting to a distracting sound. Sometimes this takes the form of focusing on one talker while simultaneously tuning out other talkers.

Many studies have shown preferential activation of brain areas responsible for these abilities in musicians compared to nonmusicians.[44] Germane to the sound mind, auditory attention and working memory skills correlate systematically with the biological processing of key sound ingredients.[45]

AUDITORY ATTENTION   **AUDITORY WORKING MEMORY**

Figure 5.5
Across the life span, musicians outperform nonmusicians on auditory attention and working memory.

Thinking: *Creativity*

Improvisation is a child of creativity. Charles Limb, a physician and musician, looked at musicians' brains as they improvised on a keyboard while inside an MRI scanner. He discovered that large areas of the frontal cortex became less active.[46] These areas typically are responsible for monitoring what we do, including behaving appropriately. Musical improvisation requires liberation from our conscious scrutiny. It is, however, also based on hours and hours of deliberate, conscious practice ahead of time. Herbie Hancock says his life experience informs the choices he makes in his music compositions but notes that "*how* it gets expressed is often a complete surprise."[47]

Making music is arguably one of the best ways to foster cognitive strengths such as attention, working memory, and creativity. Remarkably these strengths are not just musical, but transfer to other activities, most notably speech.

Music Medicine

In his book *Healing Songs*,[48] Ted Gioia talks about an order of Benedictine monks in France whose health suffered when an edict from Vatican II ordered them to stop their chanting. They became listless, irritable, and chronically exhausted. Their physical health was compromised as well, with disease rates skyrocketing. After chanting was reinstated, their health and well-being returned.

Involuntary motor tics in Tourette's syndrome can be suppressed during music making.[49] Oliver Sacks talks about watching a drum circle of people with Tourette's. After a rather chaotic start, the involuntary, unsynchronized movements of the participants eventually synchronized into a well-coordinated rhythm, as though their nervous systems were bonding with each other.[50] Country music singer Mel Tillis stuttered when he talked but not when he sang.

These anecdotes demonstrate the connection between music and health—both mental and physical—and this connection has

Figure 5.6
Logo for 2018 Music Medicine conference hosted at Northwestern University.

a history that spans ancient times to the present.[51] Music medicine is an enormous topic that goes well beyond the scope of this book. Music is increasingly entering mainstream medicine.[52] It has been pressed into service in the treatment of traumatic brain injury,[53] for mitigating stress in war and disaster victims[54] and addressing the stress that comes with an intractable illness.[55] It can mitigate memory loss in dementia.[56] It can strengthen language skills in children with autism[57] and other children with language delays or reading difficulties.[58] Music is an effective therapy for movement disorders such as Parkinson's disease,[59] stroke,[60] and difficulty with respiration, swallowing, and speaking.[61] Music can train children with hearing loss to better understand speech and make use of speech prosody.[62] These wide-ranging applications are summarized in the logo (figure 5.6) of a recent Music Medicine conference at Northwestern University, which you can view in its entirety from the Brainvolts website.

Music medicine draws on our sound minds' connection to how we move, think, sense, and feel. Through the privileged connections between the sound mind and these vital brain functions, music can provide a powerful form of healing. Music is an undertapped resource with enormous potential for growth in health care. The sound mind is at its heart.

6
Rhythm: Inside and Outside the Head

If sound loses time, it loses meaning.

My husband reads to me every night in bed before we go to sleep. With our special bear, Oatmeal, tucked in between us and listening too, this is a wonderful way to end the day and is a quotidian highlight. We deliberately choose books that are familiar—oft-read children's classics by E. B. White and the Harry Potter series make frequent appearances—so I do not worry about missing something important when I drift off. I have noticed that after some time—it can be as little as a few minutes if I am especially tired—the meanings of the words are gradually eclipsed by the sounds. I begin to hear sounds and rhythms instead of words and story, and the waxing and waning of the accents and stress patterns become a calming, lulling, treasured experience that soothes and resets me after a long day.

Why do we care about rhythm? It connects us to the world. It plays a role in listening, in language, in understanding speech in noisy places, in walking, and even in our feelings toward one another.

Rhythm is much more than a component of music. Nevertheless, music is probably what first comes to mind when we hear the word *rhythm*: drumming, jazz, rock and roll, marching bands, street performers with wooden spoons and five-gallon buckets, drum circles, time signatures, stomp-stomp-clap—we will, we will rock you—adventures on the dance floor, beatboxing, incantations, mantras, and prayers. Beyond music, we experience the rhythmic changes of the seasons. Some of us have menstrual cycles. We have circadian rhythms—daily cycles of mental and physical peaks and troughs. Frogs croak rhythmically to attract mates and change their rhythm to signal aggression. Tides, seventeen-year cicadas, lunar phases, perigees, and apogees are other naturally occurring rhythms. Human-made rhythms include the built world—street grids, traffic lights, crop fields, mowed designs in baseball diamond outfields, the backsplash behind the kitchen counter, spatial patterns in geometric visual artforms.

Maintaining rhythm is almost a biological imperative for some of us. My musician husband can get infuriated if we are playing together and I stop in the middle of a song. He needs to keep the pulse going. I have my own hiking version of this imperative. I need to *keep moving* no matter how tired I get, one foot in front of the other, even if I slow *way* down, or essential energy just drains away.

Music and rhythm are rooted in every known culture.[1] What parent does not use rhythmic rocking to soothe a crying baby? The repetitive sounds and silences that comprise rhythmic patterns make dancing possible, aid in the memory and reproduction of music, and facilitate group singing, playing, or drumming. Rhythm has been used for millennia to tie societal members together—the chants of a religious order or the cadence calls of military ranks are just two examples. Poetic works thousands of years ago, such as those of Homer, were chanted or sung with rhythm serving a mnemonic function.[2] Repetitive or complex work engenders rhythmic accompaniment, in some cases to break the monotony, in others to actually help you perform the work better. Workers performing hard labor like rock breaking chant to keep their sledgehammers swinging in rhythm.[3] Postal workers in Ghana hand-cancel stamps with a distinct rhythm.[4] Rug weavers in Iran use chants with a complex musical structure to communicate weaving patterns to their co-weavers.[5] All musical systems and styles have organizational rhythmic motifs. Indeed, the very universality of rhythm is a strong argument for the existence of biological processes governing the perception and production of rhythm.[6] Rhythms in the brain have been called out as a basis for consciousness itself.[7]

Language probably does *not* immediately come to mind when we think of rhythm. You might have had a high school literature class where you learned about prosodic feet—iambs, trochees, and anapests. But outside the context of poetry, we rarely think about speech having a particular rhythm. After all, we are likely to say "Oy Bill—you ready yet?" Not, "Hey there Bill,/do you think/it is now/ time to go?" so that it conforms to dactylic tetrameter. What about rhythm and reading? Here, too, we are unlikely to associate rhythm to reading unless we are reading poetry. In fact, *rhythm is a necessary ingredient of linguistic communication* itself.

Rhythms Fast and Slow

Rhythm can be viewed through the lens of shorter and longer time scales. Speech has phoneme-, syllable-, word-, and sentence-length rhythmic units, each unfolding at their own rate. We understand that speech comes in different sized units—the sound an individual letter makes, the phoneme, at one extreme, and the slowly rising and falling loudness and pitch contours that unfold over the course of a sentence or group of thoughts on the other. This latter one is the nighttime reading rhythm I fall asleep to. These entwined elements of speech constitute rhythms that must be sorted by our sound minds. We can try to focus on the slow parts of speech (say, the fluctuating pitch of the voice) and ignore the fast (the vowel and consonant sounds that convey the meaning of the words) or vice versa. But this is usually not possible and rarely desirable.

This temporal hierarchy is at work in music too. Music is a mix of slow phrases, steady beats, sustained notes, rapidly changing notes, trills, and drum crashes. Entwined temporal structures are in environmental sounds as well—when walking through the woods, we simultaneously hear slow footsteps, the unfolding crunch of leaves underfoot, and the rapid snap of a twig. Much as sound units come in different lengths, brain rhythms come in different speeds.

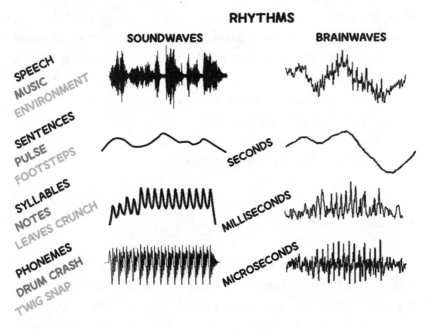

Figure 6.1

Sound waves and brain waves occur over fast and slow rhythmic time scales (top row). A sound has a broad overall shape; it turns on and turns off. This slow ramp, measured in seconds, is shown in the left, second row. Zooming in, we see the repeating waves that define its pitch (left, third row). The pitch of adult speech ranges from about 80 to 250 Hz, which equates to tens of milliseconds. Further increasing the magnification are the vowels and consonants with frequencies up to the thousands of Hz, or in the microsecond time scale (left, fourth row). Each of these time scales (seconds, milliseconds, microseconds) is maintained simultaneously in the brain (right column).

Subcortical structures are equipped for microsecond timing while the cortex is better suited to integrating sounds over a longer time scale.

Brain rhythms can be measured both when at rest and when performing an activity. When listening to speech, there are fast brain rhythms that entrain to the fast phonemes, the near instantaneous consonant sounds. Middle-range rhythms in the brain track the rate of syllables. Slower brain rhythms correspond to the slow oscillations of phrases and sentences.[8]* Similar nested brain patterns are active when listening to music.

Rhythm is *in* Us

We've all heard him. The beginning piano player. The "Eensie Weensie Spider" and "Frère Jacques" plinking hesitantly out of the piano. Usually getting the notes right is paramount to a beginner; hitting the right keys trumps hitting them at the right time. It is heartwarming to listen as our child plays (mostly) correct notes at wildly incorrect times. What goes on in the brain when we hear music being played on beat or off beat?

Imagine a metronome ticking at about 144 beats per minute (bpm). Popular songs in this range include Blondie's "Call Me," the Beatles' "Back in the USSR," and the Rolling Stones' "(I Can't Get No) Satisfaction." It's a fast, *allegro* rate. Measured another way, these songs have about a half second between their beats. If we play a conga drum by itself at this rate and record brain waves to it, we will see neural activity repeating every half second (boom, boom, boom, boom, or "one, two, three, four"). But what does the brain do if you listen to the conga drumming along to a song that

---

*These brain rhythms are named with Greek letters. Exact frequency cutoffs are somewhat fluid, but roughly, the slowest are delta (1–4 Hz) and theta (4–8 Hz) and the fastest is gamma (30–70 Hz), with alpha and beta falling in between. These ranges span sentence-slow to phoneme-fast.

matches this beat? The brain produces a new rhythm! In addition to a response peak every half second (where musically speaking the "ones" are), you see another, smaller peak halfway in between (1 *and* 2 *and* 3 *and* 4 *and*; "FLEW in FROM mi-AM- i BEACH"). The brain has worked out the strong/weak pairs comprising the song's meter. This tells us the brain entrains and reinforces both explicit and *implied* rhythms in the music.[9] This extra rhythm in the brain wave does not occur when the song is deliberately misaligned with the conga beat. A similar example of the brain creating a beat comes from Brainvolts alum Kimi Lee, who found that the fundamental frequency of an identical speech sound is enhanced when it occurs on the "1" in a four-beat sequence.[10] The sound mind's response to a drumbeat is deeply shaped by its aural context. Rhythmic organization operates automatically when we listen to sound. If our rhythmic expectations are violated, our brains behave in a different manner because of our inherent internal sense of rhythm.

Rhythm Intelligences

Imagine the familiar rhythm "Shave and a haircut, two bits" and tap it out on a table with your finger. Did you tap seven times? Now imagine it again and tap your foot to it. Did you tap seven times again? Or fewer? For me, when I tap my finger on a table, I tap to every note (ignoring the rests). When I tap my foot or snap my finger along to music, I typically tap or snap to the beat (or pulse) of the song, not every note. When I tap my finger on the table, hitting the "sounds" and ignoring the "silences," I am tapping out the *rhythmic pattern*—I am keeping track of how long or short each note is and where the pauses occur. When I tap my foot, I tap four times, to the underlying *beat or pulse* (figure 6.2), which in this example includes a silent beat. Music has both a pulse and a rhythm pattern, notated by time signature and note/rest durations, respectively.

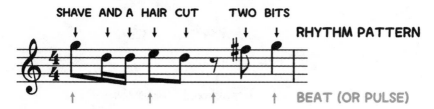

Figure 6.2

The rhythm pattern of "Shave and a haircut, two bits" is defined by the duration of the notes and rests that make up the tune (top row of arrows). In musical notation, pulse is signified by the time signature, in this case 4/4. The bottom row of arrows depicts the four beats, which may occur during either a note or a rest. Can you simultaneously tap your foot to the beat and your finger to the rhythm pattern?

Before I began studying rhythm, if you had asked me about the skills involved in tapping out rhythmic patterns versus tapping out the beat, I would have said, "You're probably either good at both or not so good at either." If someone can tap to a beat, she can also tap out a rhythm pattern, right?

Wrong. There are *multiple rhythmic intelligences*. You cannot predict how someone will perform one rhythm task by how they perform a different rhythm task. This was first noticed in extreme cases where a person with brain damage could be impaired in one sort of rhythmic ability but not another.[11] We have since learned these distinctions are fundamental to how the system works: we see dissociations between rhythm skills in all of us,[12] confirming the idea that "rhythm" is not an all-or-nothing ability, and more intriguingly, our proficiency executing one type of rhythm or another bears on our language skills. Both beat-keeping and rhythm-pattern skills predict language development and reading ability;[13] however, only rhythm-pattern ability has a bearing on understanding speech in noise[14] as we'll see shortly.

Brain Rhythms

Rhythm pattern skills are associated with slower brain rhythms (seconds), while beat-keeping skills are associated with faster brain

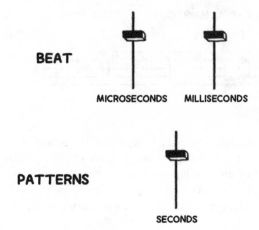

Figure 6.3
Drumming to the *beat* corresponds with sound and brain rhythms in the micro-to-millisecond time range. Drumming rhythm *patterns* correspond to sound and brain rhythms in the slower second range.

rhythms (milliseconds and microseconds; figure 6.3).[15] Phonemes, syllables, and sentences range from microsecond to millisecond to second timing, respectively. Brain rhythms can predict language development in infants and children.[16] Brain rhythms can also determine a person's strengths and bottlenecks related to language and the ability to make sense of an auditory scene while listening in noise.

Rhythm, Language, and Listening

Rhythm is tied to language. Children who recognize differences in rhythm patterns and tap to a beat learn to read and spell more easily.[17] Several beat-keeping skills are impaired in older children with dyslexia.[18] We have found a link between beat keeping and language development in adolescents[19] and in children as young as three years old.[20] What is the connection between rhythm skills and what might appear to be unrelated skills like reading and writing?

There really *is* rhythm in language, beyond the rhyming of poetry. It is inherently a part of pronunciation. Rhythm matters, even in single words. "Record," "contrast," "project," and "produce" can be either nouns or verbs depending on which syllable is stressed. Running speech also has rhythm. A YouTube search for "drumming to speech" will uncover some nice examples; a personal favorite is the one with the scene from the Gene Wilder *Willy Wonka* movie. The video shows a drummer playing along to the rhythms of the dialogue between Willy and Grandpa Joe, so you cannot miss the rhythm in speech. Tabla player Zakir Hussain tells us his father taught him to speak using drum rhythms when he was a baby. In tabla, each finger is assigned a syllable, and playing the tabla is akin to speaking in phrases. In all languages there is a definite rhythmic aspect to spoken language, brought about by alterations in stress, duration, and pitch of the syllables. This was resoundingly brought home to me firsthand when Zakir accompanied me on the congas during a speech on rhythm and language. Quite simply, rhythm in speech tells us when important information starts and stops. Stressed syllables emerge at roughly regular intervals and, importantly, carry the majority of the information of speech. With an ongoing rhythmic flow, the listener is guided to the important features of the sentence by the expectancy rhythm sets up and, so primed, we understand the content of the spoken word better.[21] With the understanding of spoken speech comes the ability, when learning to read, to make the necessary connections between the sounds of language and its written form.

One of the greatest impediments to successful spoken-word communication is noise. The rhythm of speech helps us. This is because the rhythm in speech helps us fill in the gaps when noise causes us to miss a few words. Just as a rhythm pattern evolves over the course of a measure of music, running speech evolves over time and thus is suited for a slower scale of auditory processing. The strong and weak stresses, phrases, and boundaries of language are relevant to the whole spoken sequence. The ability to reproduce rhythm patterns seems to draw on the same skills required for forming the auditory scenes that comprise hearing speech that is barely audible above noise.

Hearing speech in noise can be partly predicted by one's ability to tap out rhythm patterns.[22] And the better you are at navigating rhythms—and musicians of any stripe (not just drummers) fall into this category—the more you can capitalize on the rhythmic patterns in speech and eke out what was said despite the noise.[23]

Rhythm and Vocal Learning

Have you heard about Snowball? If you haven't, stop reading right now and search for "Snowball the cockatoo" on YouTube. Snowball the sulfur-crested cockatoo dances along to pop music. There is no mistaking it. Snowball bobs his head and steps his feet in time to the rhythm of Michael Jackson, Lady Gaga, and, most famously, the Backstreet Boys. John Iversen and Ani Patel studied Snowball's dancing.[24] By systematically varying tempos and observing corresponding shifts in Snowball's movement, they verified that Snowball was reacting to the beat. This is in contrast to "dancing" horses, for example, who do not react to music but whose rhythmic steps are cued by the rider. So Snowball's dancing, in addition to being seriously cool, raises some questions. Who else can do this? Can other bird species dance? Any other animals? Why can't my dog—he's smart? Chimps must be able to, right? They are closer to humans than a cockatoo. It turns out Snowball is part of a small and select group of animals that can keep a beat. To date, beat keeping is confirmed only in a variety of birds including parrots and cockatoos, sea lions, elephants, and humans. That's it.

What does this seemingly disparate group of animals have in common? Along with bats, whales, seals, hummingbirds, and songbirds, they are "vocal learners." This means these species have the ability to *imitate* new sounds they hear. Most animals, however intelligent, cannot imitate sounds. Take your dog, for example. She may know a dozen or more words. I have never met a dog who does not understand the word "walk." And yet, however deep the associations in her mind between certain words and what they represent, she will never *say* "walk." Dogs, along with most of the animal kingdom, are limited to a handful of vocalizations. Vocal learners, however, can move beyond their innate sounds. Parrots can "speak." Songbirds, which we will revisit later, learn their songs via imitation; a bird raised away from others of his species will not develop the same song as his peers but rather an impoverished and unstructured aberrant deviation. Human speech, of course, is a testament to our species being masters of vocal learning. These imitative abilities arise from extensive connections in the brain between the auditory and motor regions that are missing in most other species. A by-product of this connectivity is the ability to *predict the timing of future beats*. This is the key to Snowball's and humans' beat keeping. We do not simply react, like a horse or a chimpanzee, to a present or past cue; we anticipate future beats and can move our bodies accordingly.

Rhythm and Movement

Sound *is* motion, the motion of air. We have spoken about rhythm thus far as something that is heard. However, the flip side is movement. You cannot hear a drum unless someone beats it. You cannot snap your fingers along to a song unless you, well, move your fingers. You cannot produce speech without moving your mouth. Moving is intertwined with hearing when it comes to making and listening to music and speech. When just *listening* to speech, or even *imagining* a singer, the brain areas devoted to the movement of our mouths are activated.[25] Likewise, listening to a piano melody activates the parts of the motor system involved in controlling fingers in people who know how to play the tune on the piano.[26] Your sound mind "moves" to music even if you are quiet as a statue, especially if it is a piece of music you have played.

If you are walking down the street talking to a friend, it is likely the two of you unconsciously synchronize your footsteps.[27] This synchronization helps us communicate—the number of footfalls is halved, thus decreasing the chance that any given speech sound is drowned out by a footstep. Not only can you hear your friend better but, if you and your friend happen to be a pair of wild animals, you are better able to monitor and detect nearby prey or predators.

Infants just days old attend to rhythms,[28] but what determines which rhythms they choose to listen to? It turns out the *motor* component of rhythm plays a role in preferences. In one study, seven-month-old babies were exposed to an ambiguous rhythm pattern.[29] The ambiguity arose because one could ascribe either a 2/4 or a 3/4 time signature to it. That is, you could reasonably count out either 1-2-1-2-1-2 or 1-2-3-1-2-3. Experimenters bounced babies at *one* of the two possible meters—baby May was bounced on every second beat and baby June was bounced on every third beat. Then later, without being bounced, each heard *stressed* versions of the same two rhythms, meaning this time the location of the 1's was obvious. May preferred the 2/4 stressed one and June preferred the 3/4 one, as measured by how long they listened to each before turning away. Rhythm preferences are formed early in life. Notably, babies did not form a preference for 2/4 or a 3/4 meter when they simply *observed* someone else bouncing; their own bodies had to move to seal the deal.

Rhythm and Socialization

How we feel about another person is conveyed by rhythm. Walkers synchronize their steps to aid communication. Snowball will dance along with you, but if you dance out of sync with the beat, he will turn away from you. Social encounters with rhythm influence our attitudes. The extent to which a person synchronizes with an experimenter affects the person's opinion of the likability of the experimenter. University students were instructed to tap along to a metronome while an experimenter was also tapping their finger nearby. When the experimenter tapped at the same rate, the rating given in response to "how likable was the experimenter?" was higher.[30] Likability aside, the mere presence of another person drumming along with a task will improve performance. Pre-school-age children asked to perform a rhythm synchronization task perform it better if they are drumming along with another human than to an impersonal beat coming from a loudspeaker.[31]

Even in very young children, being (literally) "in sync" with another person engenders positive feelings toward them. An experimenter bounced fourteen-month-old children along to music either on the beat or intentionally off the beat. When the bouncing session was over, the baby was placed on the floor and the experimenter deliberately dropped an object and acted out needing help to pick it up. The babies who were bounced on-beat were much more likely to help the experimenter retrieve the object, having apparently formed a social bond, via rhythm, that prompted cooperation. The off-beat babies were less likely to help.[32] *Rhythmic* synchrony had led to *interpersonal* synchrony.

Along similar lines, the brain rhythms of musical performers and their audiences have been measured in concert settings. The brain rhythms tend to synchronize, and the more synchronization between performer and listener, the more listeners report enjoying the performance.[33]

Music in general, and rhythm in particular, does an uncommonly good job fostering a sense of community. Indeed, music being played at negotiation sessions helps to smooth the conversations and leads to breakthroughs and compromises. Musicians Without Borders is used to form relationships in troubled regions around the world, to bring hope, comfort, and healing to diverse populations.[34] The

Resonance Project and the Jerusalem Youth Chorus, which are forming bonds between Israeli and Palestinian children, are other examples of using musical rhythm to overcome differences. The early days of the 2020 coronavirus pandemic were marked, in some European countries, by daily sessions of songs sung from balconies to connect with others at a time of isolation and to communicate appreciation and solidarity with health-care workers.

Rhythm for Health

Traditional healers in all regions of the world have relied on rhythm as a primary force in their rituals and practice.[35] Today, rhythm helps us exercise as we move to keep ourselves healthy.[36] Therapists have long used our capacity to perceive sound patterns to strengthen communication skills. They rely on rhythm and the concepts of entrainment to a beat, violations of a beat, and pattern recognition as core features of their protocols,[37] reminiscent of the scene in the Colin Firth film *The King's Speech*, where King George VI overcame a stuttering problem by rhythmically singing his words. Rhythm capitalizes on our sound minds' auditory-motor connection.

First mentioned by the American Medical Association in 1914, music therapy was put to work helping wounded World War I soldiers recover from their injuries, including what we now call traumatic brain injury. Rhythm-based therapy has a growing status in recovery from concussion and other brain injuries, addressing both cognitive and emotional health.[38] Rhythm is used to great effect to pace walking in individuals with movement disorders such as Parkinson's disease.[39] After all, walking *is* a rhythm. Other disorders that involve movement, such as aphasia, stuttering, difficulty with respiration, swallowing, and speaking, respond to music therapy.[40]

Therapy involving rhythm also has shown promise in addressing communication and social behavior in people on the autism spectrum.[41] Children who cannot otherwise speak can form words and sentences when accompanied by a clear rhythm. There are children on the autism spectrum who will not engage in a verbal conversation but will gladly carry on a *rhythmic* conversation with another person on drums. And, moving in synchrony positively affects how we feel about each other.[42]

If I had a magic wand, I would make rhythm an indisputable part of language therapies through music and rhythm-based instruction. This would mean a closer alignment among fields of speech therapy, music, and music therapy. There are explicit rhythm-based training programs which make synchronizing to rhythm an explicit core exercise with the aim of improving timing in the brain. Some have been used to bolster language, reading, and communication skills, and do so with tasks that engage both the slow and fast sound processing circuits in the brain, thus drawing on multiple rhythm intelligences.[43]

Music with a regular and predictable rhythm can lead to states of enjoyment or emotional transcendence.[44] Pythagoras viewed music as a gateway to the realm of the dead, at least judging from his supposed dying request that the monochord, an ancient one-stringed instrument, be played during his final moments. Gregorian chants have been described as "so rich in overtones that you have the impression they are angels, not men."[45] Grateful Dead drummer Mickey Hart and I have discussed the calm yet alert and energized state that drone compositions—musical pieces consisting of sustained sounds produced by monochords or other instruments and manipulated in the studio to swell and build—can induce. We are working together to investigate the neurophysiological reaction to some of his drone compositions.

A while back, one of my sons got a hairline fracture in his foot. As he was not healing as fast as his physical therapist was hoping, he was assigned daily sessions with a bone vibrator. The idea behind vibration therapy is that if you cannot use your musculoskeletal system normally, for example due to an injury or osteoporosis, you miss out on the natural stimulation that occurs as your muscles imperceptibly relax and contract to maintain posture. This can lead to bone tissue atrophy. Imposing vibrations at around 30–50 Hz at the injury site simulates natural postural adjustments, stops the reabsorption of bone tissue, and promotes the bone growth that would ordinarily be achieved as part of typical day-to-day movement.[46] It appears that low frequency vibrations spur activity in the stem cells that make cartilage, muscles, and bones. This process may also be useful for strength training in noninjured people.

It turns out the vibration rate of a cat purr is in the exact same range as used in vibration therapy for bone growth. Cats purr when they're happy, of course, but in what other circumstance do they purr? When they are injured! There is a hypothesis that cats purr as a mechanism to keep their bones and muscles stimulated and healthy and to restore their health when injured.[47] Maybe it is not a coincidence that cats have better bone health and a lower incidence of osteoporosis than dogs. Maybe this is the secret to their nine lives.

Conclusion

Why do we care about rhythm? Because *sound is motion and sound moves us*. The auditory and motor systems combine to enable us to communicate. When we engage with rhythm, whether a precise in-the-moment beat or a longer rhythm sequence, we rely on precise timing. Our very brains are wired for rhythm across these time scales. The currency of the nervous system—electricity—is nothing if not rhythmic. To my mind, the crackling of action potentials as they respond to sound is closer to a one-to-one correspondence of stimulus to response than in any other sensory modality. Physiologists commonly capitalize on this and literally "listen to the brain" to help guide their electrode placement during their experiments by playing the crackle of neural activity through a speaker. I love to listen to the brain's language of timed, rhythmic (and arrhythmic) electrical impulses. The better we understand the biological basis of rhythm, the better we will be able to employ rhythm—in all its guises—to improve communication and to better understand ourselves.

7

The Root of Language Is Sound

Sound + Learning = Language

—Kasia Bieszczad

If every time the word "ball" was spoken, it was pronounced differently, and every time it was written, it was spelled differently, no one would ever learn to read or understand the word "ball." Language relies on consistency. When learning to speak, a child has to hear the word "ball" in reference to the round rubber orb in her hands over and over again to make the sound-to-meaning connection between the word and the object. With respect to reading, consistency comes in at least two flavors. First, we rely on a reasonable consistency between the sounds of language and the orthographical translation (written representation) of these sounds.* Letters connect us to the sounds of language. The process of "sounding it out" would be meaningless if there were not a reasonably consistent

---

*Phonetic languages use letters to represent sounds. The earliest known writing system linking symbols to sounds is the Phoenician alphabet dating to the eleventh century BCE.

mapping between the letters and the sounds they represent. Second, we rely on a consistent auditory brain to help us make the sound-to-letter connections.

In most languages, spelling bees do not exist.* There is a near one-letter-per-sound mapping in many languages. If you hear a word in Spanish or Italian or Russian or Finnish, you can probably spell it correctly on the first try. You rarely have to puzzle through the "is it a *c* or a *k* or a *ck* or a *ch* or a *qu*" sort of conundrum that English spelling brings up.

In English, with its borrowing from Greek, Latin, French, German, and other languages, we have a hodgepodge of sometimes arbitrary sound-to-letter mappings to keep track of and memorize. Another source of the sometimes-capricious lack of consistency in spelling is the Great Vowel Shift that took place in England in the fifteenth and sixteenth centuries. Prior to this time, there was more consistency between sounds and letters. As in French, an English "i" was reliably pronounced "ee." So "bite" was pronounced "beet." The "ou" as in "house" was pronounced "oo" like "moose."† Pronunciations gradually changed but spellings retained their pre-shift form, leading to the current situation where the forty-odd sounds (phonemes) of English are represented by an astounding 1,120 different letter combinations.[1] Many have heard the old joke that fish should be spelled "ghoti": *gh* as in laugh, *o* as in women, and *ti* as in nation. In contrast, my other language, Italian, has fewer sounds (twenty-five) but only thirty-three different letters or letter combinations that produce them. The extent to which sounds do or do not map directly to letters is known as orthographic depth, and English is one of the deepest. Indeed, English-speaking children,

---

*To be fair, it is my understanding that spelling bees are a uniquely *American* thing. They are virtually unknown in Britain and other English-speaking countries.

†One theory behind the Great Vowel Shift is that anti-French sentiments were common among Britons in the Middle Ages, and this was a way to further distinguish the sounds of English from the sounds of French.

along with French, Danish, and some other orthographically deep language speakers, lag behind their shallow-orthographic peers in reading acquisition.[2] Common to all languages is the need to make sense of sound to read.[3]*

There is also a need for *heard* consistency. A ten-year-old—let's call him Danny—came to Brainvolts some years ago. Jane Hornickel was a graduate student at that time whose interest was sound processing in dyslexia. Danny was bright—IQ testing verified this—but he was failing in school. His reading was slow and labored. He had difficulty breaking speech into its component parts (sounding it out) and lacked fluency. Ultimately his comprehension suffered. Once he moved from the "learn to read" to the "read to learn" phase of his education, there was trouble. Everybody—parents, teachers, peers—could see Danny was a smart, engaging, and engaged child . . . who simply couldn't read. But Jane was able to see something else—*inconsistency in his neural processing of sound.*

When you hear a sound, the brain fires with a certain signature pattern. We can measure this electrical pattern with scalp electrodes. When you hear the same sound a second time, the brain pattern should be the same. Jane discovered in Danny's case, the consistency was just not there. It was as if the sound, at least as far as Danny's sound mind was concerned, was slightly different every time he heard it. How could Danny be expected to make the sound-to-letter and letter-to-sound connections that make fluent reading possible if there was no consistency in how his brain was hearing the sounds?

Jane had an idea about how Danny might be helped to overcome the challenges posed by his dyslexia. But before we get to that, what

---

*Children who are slow readers in *any* language have much in common with their English-speaking counterparts. They have similar problems in the speed of their reading and in sounding out words. There are also commonalities in brain function in dyslexics across languages.

do we know about the sound-reading connection? How much does sound really matter to reading?

Sound and the Reading Brain

There is no reading center in the brain. "Human beings were never born to read," writes Maryanne Wolf.[4] We have only been reading for a few thousand years—evolution does not work quite that fast. Maybe our distant descendants will have reading centers in their brains, but to the best of our knowledge, twenty-first century humans do not.* Yet we *do* read. We accomplish it by coopting other parts of the brain, most notably the sound mind. The visual brain is involved, too, of course.[5] But auditory areas, including those that govern both speaking and understanding spoken language, play an outsize role.

I'm often asked, "What does sound have to do with reading?" The connection between sound and reading is not immediately obvious. We generally read in silence. Yet language is rooted in sound, and reading is rooted in language. Reading aloud explicitly connects sound with written language. When we learn to read, we must connect the sounds and sound patterns of the language we speak with the letters they represent. Poor readers struggle with sound,† and

---

*Nor did they in the fourth century BCE, when Plato looked upon the printed word with skepticism, worrying that it got in the way of memory. "If men learn [writing], it will implant forgetfulness in their souls; they will cease to exercise memory because they rely on that which is written, calling things to remembrance no longer from within themselves, but by means of external marks" (Plato, *Phaedrus* [Indianapolis: Hackett, 1995]).

†There is no denying reading engages our vision (or tactile sense, in the case of Braille). A contributing factor in dyslexia is a deficit with motion and timing in vision, in contrast to color and spatial perception. Eye strain or visual distortions occur in dyslexics at a higher rate than in the general public. Yet despite reading's obvious link with vision, sound processing seems to make an especially large contribution to reading.

auditory processing repeatedly reveals itself as one of the biggest bottlenecks when it comes to challenges in learning to read.[6]

Language learning depends on discerning *sound patterns*. It is natural for us to hear a sentence and know where one word ends and the next begins. But consider that, acoustically, there are no overt gaps between words. Phonemes blend into syllables blend into words. Silences between words are no longer—and are often shorter—than silences within a word in running speech. There are clues we learn to help us. For example, the letter/sound combination "mt" rarely occurs within an English word. So if we hear a snippet of speech that contains "Sam took," we intuit that we haven't encountered a new word "samtook." We learn these tricks of the English-language trade very early—already at two days of age![7] University of Wisconsin professor Jenny Saffran discovered that eight-month-olds can learn sound rules of a made-up language after a mere two-minute exposure.[8]

Pattern learning is evident in the neural processing of sound. Brainvolts graduate student Erika Skoe found neural enhancement of harmonics once the pattern of a made-up language became familiar.[9] Similarly, harmonic enhancement is evident when a speech syllable occurs in a regular sequence versus occurring randomly in a string of different syllables.[10] Children with language problems, however, cannot learn to pull these implicit rules out of language.[11] Children with hearing loss likewise have difficulty with pattern-forming language tasks,[12] and children with autism exhibit distinctive patterns of brain activity while being exposed to this sort of artificial language.[13] Enrichment in the form of bilingualism and music training, on the other hand, enhances the processing of sound patterns.[14]

There is yet more evidence that sound is part and parcel of language. One might predict a musician would be good at discriminating between a pair of very close pitches, say 1000 Hz and 1003 Hz. And that prediction would be accurate.[15] But it is less self-evident that telling pitches apart (frequency discrimination) would have any relationship with reading ability. Yet a disproportionate number of dyslexics, both children and adults, struggle to discriminate pitch pairs,[16] pitch patterns,[17] or dynamic moving pitches (i.e., FM sweeps).[18] This reduced ability to distinguish *sound ingredients* is independent of intelligence and is evident in how the brain responds to sound.[19]

Another ingredient that poses a thorny challenge to the sound mind is timing. Sensitivity to timing is often measured by "gap detection." You play a pair of sounds—often tones or short bursts of noise—one after another. If there is enough quiet space between the two, you will hear it as two sounds, eeeee—eeeee. However, as you shorten the time gap between the two, you eventually reach a point where the gap is too short to be detected and you hear only one sound, eeeeeeeeee. People with reading impairments often need a longer gap to hear the two sounds as distinct. The sounds smudge together into one sooner (with a longer gap) than for typical readers.[20] Reading also relates to the ability to detect tones appearing immediately before a burst of noise,[21] and to detect amplitude modulation.[22] What is notable about these reading-associated hearing struggles is they can appear in *nonlanguage* sounds. That is, there is not just a link between *speech* sounds and reading but rather a link between *sound ingredients* and reading.

Children as young as a few months old can tell us a lot about what they are able to hear. Infants perceive the whole universe of language sounds, the phonemes, beats, and pitches of the world's languages. Then they lose that ability as their sound mind becomes honed to the sounds that matter to their native language.

Studies of very young children often rely on their eagerness to view engaging objects. By using the appearance of a dancing toy bear as a reward, babies can be taught to identify a change in a sound sequence. When incorrect, no dancing bear appears. April Benasich at Rutgers University capitalized on this to explore the role of sound in language development. First, she looked at seven-month-old babies' performance on this task. Next, she retested them years later, at age three, and looked at their language outcomes with respect to their seven-month results. The seven-month results were strikingly predictive of their age-three language comprehension, expression, and verbal reasoning abilities. In similar studies, distinguishing nonspeech sounds at prereading ages predict later phonological awareness and reading abilities.[23] Moreover, babies from families with a history of language impairments perform worse on the sound processing task, suggesting a hereditary component.[24]

Years ago I participated in a think tank on the topic of language and the brain at the Santa Fe Institute. There, I witnessed Michael Merzenich and Paula Tallal combine their complementary scientific approaches for social good. A pioneer of brain plasticity, Mike had demonstrated sensory and motor brain changes with experience—good and bad. Paula, a professor at Rutgers, had discovered some children with language disorders could not distinguish sounds that form the building blocks of speech. They soon published two landmark studies, demonstrating that after performing a sound training regimen, school-age children improved on a variety of language tasks.[25]

This finding, and others like it, has motivated an industry to get auditory training materials into the hands of schools and parents to address language, reading, and learning difficulties. Merzenich and Tallal went on to found a company that produced sound-based training games. Brain changes can accompany language gains following these "brain training" games,[26] and some public school systems in the US and Canada have implemented these training exercises, with subsequent academic gains reported. April, meanwhile, noted a sharpening of auditory brain maps in babies following experience with sounds with rapid changes in frequency, such as the FM sweeps that provide the building blocks of consonants and vowels.[27] This suggests positive experience with sound can influence language outcomes. She is developing a toy to help babies zero in on elements of sound, such as the fast timing ingredients that are so important to learning the sounds of language.

A thread tying much of this work together is the dependence on precise *timing ingredients* in sound, whether it be timing distinctions,

FM sweeps, or other acoustic dimensions. In speech, this type of timing-based processing is usually the realm of the consonant. Consonants are the troublemakers in speech perception. And when it comes to people with language problems, consonants such as the distinction between "dare," "bare," and "pare" are the ones that are most problematic.[28]

That's where Brainvolts took up the gauntlet. We wanted to build upon these discoveries, but we wanted to come up with a way to use sound processing in the brain to understand how sound ingredients contribute to language. Our earliest discovery was that the brains of school-age children with language problems did not distinguish speech syllables as well as those of typical readers.[29] We already knew that processing the sounds of speech consonants is difficult for people with language disorders,[30] and now we had some biological corroboration.[31] That's when we began to dig into what the brain could tell us about *detailed processing of sound ingredients*. And along with scrutinizing granular sound ingredients, we strove for individual applicability. I wanted to move beyond thinking in terms of "poor readers" (or "bilinguals" or "musicians") as packs. I wanted to think about Johnny, Margie, and George.

The Mighty "Da"

We landed on the mighty "da." There have been refinements and variants over the years, and certainly no lack of companion syllables, words, musical notes, and environmental sounds. But there is something special in the sound of this unassuming syllable that connects it with listening, learning, and language. And it does so in a systematic way, according to its sound ingredients. It is also universal: almost every world language has a "da" sound in it. Now to shine the spotlight on the sound ingredients: *fundamental frequency, timing, harmonics, FM sweeps,* and *consistency* and how they relate to language (figure 7.1).

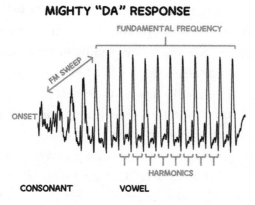

Figure 7.1

The brain's response to "da" illustrates sound ingredients in speech: timing of the onset and of the peaks during the FM sweep of the consonant, the harmonics, and the fundamental frequency.

Fundamental Frequency

If a sound is perceived as having a pitch—if it is "hummable"—the frequency we would hum is the fundamental frequency. In speech, the fundamental frequency corresponds to the speed of the openings and closings of the vocal folds set in motion by our breath. The speed of vocal fold movement is slowest for men, resulting in a deep voice (low fundamental frequency), and highest in children, resulting in a high voice. In English, the pitch of speech conveys intention and emotion—*how you meant* it, not *what you said*. Neural processing of the fundamental frequency does *not* seem to relate to reading or language development. We can cross that ingredient off the list.

Timing

We began to see the mighty "da" could uncover atypical sound processing in children with language problems when we scrutinized the brain's timing. Two Brainvolts graduate students, Jenna Cunningham and Cindy King, independently found that children with diagnosed language disorders had timing delays in the FFR to "da."[32] What's more, the delays in timing occurred in particular parts of the syllable, the response to sound onset and the FM sweep that characterizes the transition from the consonant "d" to the vowel "a." In other words, it was not a pervasive timing deficit—only the timing of the consonant was affected; we had a biological glimpse into how the sound mind could fall short in processing speech ingredients.

As time went on, findings were replicated and extended. In some cases, stressing the system by speeding up the sounds or adding background noise unveiled further timing delays.[33] In others, looking at reading skill along a continuum rather than relying on a binary diagnosis of a problem, revealed that the sound mind's relationship to language is not an either/or prospect.[34]

Harmonics

The meat, if not the potatoes, of speech is in the harmonics. Harmonics feature across the board in consonants and vowels. You turn an "oo" into an "ee" by altering the harmonics as you shape your mouth, lips, and tongue. In almost every instance where we saw a timing delay to the syllable "da" in learning- or literacy-challenged individuals, there was a corresponding reduction in the sound mind's response to its harmonics.

FM Sweeps

The trickiest part of "da" lies in the transition between the consonant "d" and vowel "a," which is defined by an FM sweep in harmonic frequencies. The identity of many speech consonants relies on harmonic bands as they shift and evolve over time. If a given band sweeps up, it is one consonant; if it sweeps down, it is another.

Children with language problems can fail to biologically distinguish syllable pairs defined by their FM sweeps.[35] This makes sense because what makes a "da" a "da" and not a "ba" or a "ga" is equal parts timing and harmonics. And this sweep of frequency over time happens very, very quickly (~1/25 of a second). This gets to the heart of why consonants are so perceptually fragile. There is just a lot going on in both *timing* and *harmonics*, and it goes on fast and concurrently. Not only are consonants difficult for people with language and learning problems, but they are among the first sounds to suffer (in anyone) when background noise is present.[36] The sound mind must work much harder to keep track of these sound ingredients as speech moves from consonant to vowel and back again. But there is also something else going on, something that ties all the ingredients together.

Consistency

I opened this chapter with a shout-out to consistency. Not a sound ingredient per se, consistency still plays an important role in the brain's encoding of the ingredients. If pitch, timing, harmonics, and FM sweeps are the ingredients, consistency is the mixing bowl (or something like that). Like Danny, whom we met above, there can be an overall reduction in sound processing consistency in the sound minds of children with learning problems. While a response to any given *single* sound presentation (trial) is more or less intact, responses are a bit dissimilar (inconsistent) from trial to trial. Some responses might be later than others; some might be smaller; some might be sharper. The neurons are firing and refiring with less *synchrony*. The individual trials do not add up into a tight, organized waveform with the microsecond precision we see in typical learners.[37] If the trial-to-trial dissimilarity is timing-based (as it often is), the resulting *sum* of all the trials will be a bit smooshed. As you can see in figure 7.2, the slower, bigger humps are fairly accurate whereas the small, fast squiggles are where most of the inconsistency shows up. This is because of a problem with the fastest microsecond timing.

Thus, several sound ingredients—timing, timbre (harmonics), and the confluence of timing and harmonics (FM sweeps)—are linked to language and reading in school-age children. Additionally, consistency in the processing of the ingredients plays a major role. But, this is but a subset of the world of sound ingredients. There is not a global

CONSISTENT

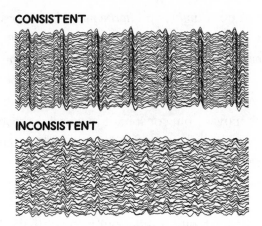

INCONSISTENT

Figure 7.2
Inconsistency is a neural signature of language impairment. Trials should line up.

movement of the faders on the mixing board; as we saw, pitch is not as involved with language. Language proficiency isn't indicated by a volume knob being turned up or down in the sound mind. Only select sound ingredients are involved, but it is a very important selection. The discovery that certain sound ingredients (depicted in figure 7.3) are not processed optimally in the brains of children with language problems furnishes a biological basis for sound's importance to language—a conceptual advance.

Using the Sound Mind to Predict Future Reading

Taking this a step further: what if we could harness the brain's response to sound ingredients to predict which child is going to have difficulty reading *before* he struggles to read? What if we could let the brain do the talking? Fueled by the discovery of a language signature in school-age children, we wanted to assess pitch, timing, harmonics, FM sweeps, and consistency in pre-readers, and then wait four or five years and check back in on their language and reading outcomes, as second and third graders. Could sound processing

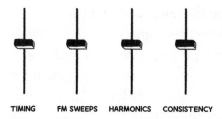

TIMING   FM SWEEPS   HARMONICS   CONSISTENCY

Figure 7.3
Timing, FM sweeps, harmonics, and the consistency of the response to sound are key ingredients for language.

in the brain at age three predict reading at age eight? Following the same child over several years can be challenging, but it is one of the most powerful scientific strategies. So the Biotots project was born. We tested hundreds of children on pre-reading, phonics, attention, memory, rhythmic skills, and a variety of auditory-brain measures. And then we did it again, year after year, for five years.

The Brainvolts team, made the experience fun for the children and built relationships with the parents. It was not unusual to get a call months ahead of a child's scheduled appointment saying, "Robby wants to come and play science games with [Brainvolts team member] Ellie." The Brainvolts team was generous with their knowledge as they addressed the questions and concerns of curious parents. Consequently, we had uncommonly little attrition, the bane of a multiyear project.

I often surprise myself by initiating long projects so at odds with my impatient nature. But the kids were adorable . . . and we learned a lot. We figured out the sound mind's reading markers we had discovered in older children—timing, harmonics, FM sweeps, consistency, but not pitch—were *retrospectively* present in three-year-olds. So our friend Jackson, who now in third grade is writing like Proust, already had robust processing of sound ingredients at age three. But his neighbor Ashlyn, who now struggles with reading at age eight, had a troubling brain signature at three.[38] Because we collected FFRs to a variety of sounds, we were able to zero in on the combination of sound and brain waves with the best predictive value. An effective sound was our tried-and-true "mighty da" syllable, with some background noise added to make things more challenging. The ingredients with the most predictive power were timing, harmonics, and consistency. Travis White-Schwoch, Brainvolts' resident statistical guru, took the lead on this fortune-telling endeavor. By creating a model combining the brain's response to these three features (a pinch of timing, a dash of harmonic encoding, a soupçon of consistency)—we achieved an astounding "hit rate."[39] We could gauge reading readiness at age three and also predict *future reading ability* when our Biotots were old enough to read.[40]

Not all language problems stem from sound processing in the brain. Equally important is knowing that sometimes sound processing is *not* the root of the problem. As a mother of boys who took their time learning to read, I would have welcomed a thirty-minute test to pinpoint or rule out a sound processing problem. If a three-year-old has an at-risk profile, parents can act early to help overcome the sound processing difficulties that hinder sound to letter to word to meaning connections that become so important in school.

Improving the Sound

Brainvolts wondered if we could see an improvement in reading and in the brain's response to sound by *improving the sound itself*. With Jane Hornickel's persistence, patience, and finesse and working in collaboration with Brainvolts collaborator and reading disability specialist, Steve Zecker, we formed a partnership with the Hyde Park Day School system, a private network of schools in Chicago that serves children with severe reading impairments. Hyde Park Day provides a course of intensive and individually focused instructional remediation with the goal of returning the child to their home school after about two years. In working with the Hyde Park Day

School network, we not only had access to a group of bright children with professional diagnoses of a learning, reading, or attention disorder but also a willing institutional partner. In contrast to the public schools we worked with in low-income neighborhoods, these private schools had every available resource to help children thrive and an administration eager for science-based strategies to help them help their students.

So what do I mean by improving sound itself? The sounds these children heard were literally improved by making sound ingredients louder, crisper, less affected by noise, and less distorted by echo. We teamed up with a hearing device company to supply children and their teachers with personal sound amplification systems (also known as *assistive listening devices*). These consisted of small in-ear devices that students wore throughout the school day, in combination with lapel microphones used by the teachers. The teacher's voice was picked up by the microphones and transmitted to the students' earbuds. In this way, each student received the same benefit—Susie in the back row could hear the teacher as well as Kevin in the front, and we could randomly withhold the amplified teacher's voice to students in the same classrooms. These students were hearing their teachers' voices in the same way they always had and, crucial for a controlled study, they were receiving the *same instruction* by the *same teachers* at the *same time* in the *same classrooms*.

We were keen to explore the scientific basis of solutions that exist in the world and not limit ourselves to ones manufactured by scientists for the purpose of an experiment. Parents and educators have access to assistive listening devices they can choose for their children. The company we partnered with to supply the listening aids needed to be willing to take a chance. It was possible we would find no biological or language benefit of their product. They participated knowing we would publish our findings regardless of the results.

We tested all the children on measures of attention, memory, learning, academic achievement, and sound processing in the brain.

Then we sat back and let the school year unfold. The children who wore the listening aids did so on average for 420 hours over the course of the school year. At the end of the academic year, we performed the same tests again.

After the school year ended, children with the listening aids showed greater improvements in their reading ability and phonological awareness (the ability to identify and manipulate the sounds of English) compared to children who had not worn the devices. Their brains' responses to speech also *became more consistent.* These biological changes were not seen in the children who completed a business-as-usual school year.[41] Moreover, the students who made the most gains in reading were the ones whose brain response consistency was the poorest at the onset of the school year, suggesting that in these most-improved cases, the root cause of their reading problem was a sound processing bottleneck addressable with intervention. I should point out the children never wore their listening devices during the brain testing. Thus, the improved sound delivery that drove making better sound-to-meaning connections fundamentally changed their sound minds. The listening device was no longer needed to maintain the sound processing gains.

We learn what we pay attention to. The children who had the teacher's voice delivered directly to their ears with clarity and adequate volume could *attend* to the lessons better. They could spend more time thinking about the the concepts of the lesson instead of figuring out what to pay attention to or which words were spoken. As more and more successful sound-to-meaning connections were made, the automatic, default network of the sound mind became better tuned to sound, as evidenced by the enhanced consistency of neural processing. So, someone like our friend Danny has options to tune his sound mind and get it responding consistently, laying the groundwork for building the sound-to-meaning connections necessary for reading fluency—emblematic of the nature of the sound mind to transform itself.

Linguistic Deprivation

A book published in the 1990s[42] famously stated that by the age of three, a child of low socioeconomic status will have heard thirty million fewer words than her wealthier neighbor. The authors posited a deficit in pre-age-three language foundation might account for the long-observed relationship between poverty and subaverage vocabulary, language development, and reading comprehension. In short, disadvantaged children entered their pre-K and kindergarten classrooms unready for school.

While the existence of a word gap has been disputed,[43] the link between socioeconomic status and language, literacy, attention, and academic achievement is not questioned.[44] And a staggering amount of research shows an impoverished upbringing can *adversely and directly affect the brain*. Childhood poverty has been linked to atypical brain structure, function, rhythm, and symmetry,[45] including reduced size of the hippocampus, amygdala, prefrontal cortex, and other brain structures important for memory, emotion, and organizing ourselves.[46]

Children from lower-income areas, on average, do less well in language- and literacy-based measures than children from more affluent neighborhoods. Early language exposure has an impact on eventual language development.[47] This can be due to a word gap, a related gap in the "quality" of language exposure,[48] noisy living conditions, or some other undetermined environmental roadblock. Whether strictly an accurate count or not, the "thirty million word gap" captured the imagination of the public and policy makers. Former President Obama, in announcing his administration's Early Learning Initiative, spoke directly of the word gap. Closing the word gap is a central plank in Too Small to Fail, a Clinton Foundation initiative to promote early brain and language development.

Municipal policies have been put in place to address this gap. A notable example is in Providence, Rhode Island, where the

"Providence Talks" program focuses on ensuring children in the birth-to-three age group are richly exposed to language before they start school. A combination of monthly in-home visits from language coaches, play groups, and wearable "word counting" technology[49] encourages expanded vocabularies and rich descriptive language from adult caregivers. So far the initiative has been successful in increasing the number of words the children hear.[50] Other US cities are looking to follow Providence's lead, including Detroit, Louisville, and Birmingham.

Brainvolts explored the *biological* impact of linguistic deprivation in children living in low-income areas. How might linguistic deprivation manifest itself in the sound mind? We looked at the brain's response to speech sounds in children from Chicago-area high schools where more than 85 percent of the student population qualified for subsidized lunch. The students were divided into two groups based on their mother's education level, as a proxy of language exposure.* All students were matched for race, ethnicity, neighborhood, age, sex, hearing, and birth history and were educated in the same classrooms. We also administered standardized tests of reading and literacy. The brain responses of the teens with mothers who had completed fewer years of formal education had a general "disorganized" look to them, with higher background noise. Moreover, encoding of the harmonics in speech was smaller, and the response consistency was lower.[51] This pattern of sound processing is evocative of the "poor reading" signature and was borne out by their reading ability. The students who had likely experienced less linguistic stimulation early in life indeed had poorer reading and literacy scores as adolescents.

---

*This method of division makes me uncomfortable. There are certainly less-formally educated mothers who bring up their children in linguistically rich environments. However, in large cumulative studies, a mother's education level has been found to be predictive of language exposure.

The neural signature of linguistic deprivation revealed two impediments: less precise processing of sound details, coupled with excessive neural noise. Enriched sonic experience through making music or speaking another language can strengthen how the sound mind processes essential sound ingredients, while overall brain health associated with physical fitness may help reduce the neural background noise. More on that later.

Autism

One of the first things parents notice is unusual or inappropriate reactions to sound. Children with autism often display an oversensitivity to sound. Or, they lack reaction to sound, especially to sounds that might be expected to induce a powerful response, such as mother's voice. In some cases, speech is delayed or absent; in others, communication is hampered by, among other things, difficulty understanding and producing the sound ingredients in speech that convey *intent*, *affect*, and *emotion*.

Individuals on the autism spectrum may have little difficulty understanding which words are spoken but may not pick up on the subtext—the emotion or nonlinguistic intent—it is delivered with. For example, they may miss anger or sarcasm. On the speech *production* end, there can be a noticeable lack of routine pitch and rhythmic changes. Instead, the speech of someone on the spectrum might seem droning and robotic, inappropriately singsong, or stressed in atypical ways. Taken together, these missed prosody cues in speech perception and production may contribute to challenges in making social connections.

The prevalence of language difficulties for people on the spectrum represents a tangible focus for efforts to help with social development. To address the question "What is going on in the brain?" Nicole Russo at Brainvolts undertook an investigation of the sound mind in children on the autism spectrum. In particular, Nicole pursued the question of prosody perception, specifically, voice pitch. In English, intonation conveys emotion (happy/sad/angry) and intent (statement/question/sarcasm). Could poor auditory processing of this ingredient of speech play a role in the challenges some people on the autism spectrum have in understanding the subtext of speech?

We created consonant-vowel syllables with intonations that made them sound like either a statement or a question. When we played them to school-age children on the autism spectrum, we often found their auditory responses did not tightly track the pitch of the syllables as seen in typically developing peers (figure 7.4).[52] In some cases, the difficulty with prosody (tone of voice) in individuals on the spectrum may thus have a sound-brain origin.

Brainvolts alumnus Dan Abrams, now at Stanford, studies the connectivity between brain regions while listening to speech. Dan discovered reduced connectivity between the auditory brain and limbic centers (responsible for emotion and reward) in children on the autism spectrum.[53] To children on the spectrum, the sounds of speech, such as mother's voice, may not scratch the same emotional itch they do in typically developing children. This fits the emerging social motivation theory of autism, which suggests emotional centers in the autistic brain are underdeveloped and so reduces the motivation for social experiences and relationships.[54] Perhaps with

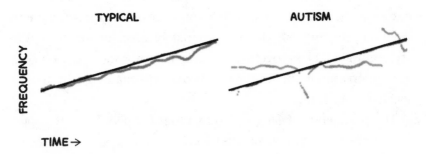

Figure 7.4
When we ask a question, the pitch of our voice goes up. The hearing brain (gray) typically tracks the pitch of a speech sound (black). The brain response does not track the pitch trajectory in autism.

autism, there is a reduction in the biological connections in the sound mind that make social interaction rewarding.

People with autism can display notable oversensitivity to sound. Researchers in Spain report a heightened response to sound as revealed by the FFR. This may reflect a breakdown in the inhibitory influences that normally keep the auditory system (notably the midbrain) in check, suggesting a biological basis for the "sensory overload" often noted in autism.[55] All of these links between sound and autism point to disruption in the vast interconnections between the sound mind with the rest of the brain courtesy of the efferent system. The links between autism and sound may lead to tailored approaches to help individuals overcome communication challenges that can be socially isolating.

Advantages of the Language-Challenged Brain

Often overlooked are the advantages and unique perspectives people with language challenges like dyslexia and autism have to offer.

Creativity can emerge in the face of language challenges. I think we all know an individual who struggles with language while excelling in other realms. I need only look as far as my middle son for an example. Reading was hard for him. When he was in first and second grade, he could observe his classmates doing this mysterious thing—reading—that he just couldn't grasp. Forget the finer points like "sounding out" and "sight words," he struggled with the very concept. What do you mean those squiggles on the page are words? Our public school's Reading Recovery program helped, as did the Bob Books series.[56] To this day, my baby—Rhodes scholar, New York City artist, founder of the Wesleyan Center for Prison Education—relies on spellcheck to prevent mistakes like "alwaze" from creeping in.

We have some less anecdotal evidence of creativity in people with language challenges. Inserting some silence just before the vowel can turn a "ba" into a "pa." Perceptually, the change from "ba" to "pa" happens abruptly. If we add a bit of silence to our ba, it still sounds like "ba." We now add some more silence—still "ba." Now a bit more—still "ba." Now a teensy bit more—bam! Now it's "pa." There is no in-between. We unambiguously hear "ba" or "pa" like a light switching on. Our sound mind forms categories it places the sounds of language into. In most people, if you play two "ba"s with a difference in the timing between the "b" and the "a," they cannot tell them apart as long as they both fall into their sound mind's "ba" category. Dyslexics, however, sometimes actually more *easily* distinguish two sounds within the "b" category than a typical reader.[57] Their sound minds, in this respect, remain *more discerning and flexible*—maintaining creative possibilities that are closed to those whose listening brains have learned to operate within immutable categories. Creativity in dyslexic people is personified by Albert Einstein, Steven Spielberg, Cher, Tommy Hilfiger, Octavia Butler, Thomas Edison, Jay Leno, Whoopi Goldberg, Ansel Adams, Andy Warhol, and Agatha Christie.

The often serious language difficulties in autism can be accompanied by dramatically overdeveloped talents in other, usually memory-based, realms. First described in the eighteenth century, these talents usually fall into one of five realms: music, art, calendar calculating, math, and mechanical or spatial skills.[58] Interestingly, although much rarer, *language-based giftedness* occasionally emerges, including extreme polyglotism and precocity in reading.[59]

Sex Differences and Language Disorders

Any schoolteacher will tell you that language problems disproportionately appear in males. In fact, a boy-to-girl ratio of more than 2:1 has been reported for reading disorders.[60] We wondered if we could discern any clues as to why this was by looking at sound processing in males and females. We were also just curious if male and female brains hear the world differently.

Sex differences in biology—aside from the obvious—show up in many realms, not least in sound. Sex differences in sound communication pervade the animal kingdom. For example, male songbirds are typically the singers, using their songs to attract females, who choose mates based on their favorite song. Similarly, male humpback whales sing to attract mates. Female birds better shift the timing of their vocalizations to evade noise.[61] The differences in the prevalence of vocalization in males vs. females raise the question of differences between the sexes in processing sound.[62] Even within a sex—for example, female mice—the hearing brain differs depending on whether the particular mouse is a mother.[63]

At Brainvolts, we studied sex differences in the processing of sound ingredients in over 500 preschoolers, adolescents, and adults.[64] Males and females differed in response timing to sound onsets—that had been known for decades.[65] However, we uncovered sex differences as well as similarities in the processing of other not previously explored ingredients of sound. These include the size of the harmonics and fundamental frequency, and the microsecond timing necessary to transition from a consonant to a vowel (FM sweeps). Preschool boys and girls process these ingredients similarly. Sex differences *emerged only later, in adolescence or adulthood* (figure 7.5). The differences may be driven by factors such as hormonal changes or life experiences. We don't know. Other measures, such as the consistency of the response and the amount of background neural noise, do not differ between the sexes at any age.[66]

Sex differences in sound processing contribute to our understanding of why males seem more vulnerable to language disorders than females. Across the board, where male and female responses diverge, the male response is the poorer one, either smaller or later, suggesting a possible biological liability when it comes to processing language. Notably, the sex differences were seen for FM sweeps and harmonics, the very sound ingredients that, along with consistency, best track with language ability. What purpose do these sex

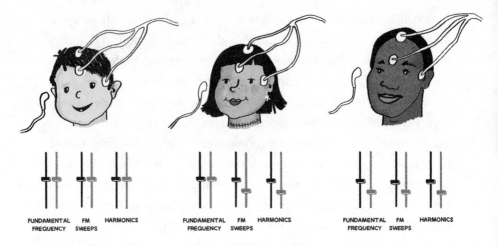

Figure 7.5
Sex differences emerge with age. Three sound ingredients are similar in preschool girls and boys. By adolescence, FM sweeps and harmonics diverge. In adults, men differ from women on all three ingredients. There are no sex differences in consistency and background neural noise level at any age. Black faders are female; gray faders are male.

differences serve in humans? It would not surprise me if these small but reliable hearing differences will one day prove to be important for how we communicate or for reasons we have yet to discover.

Strengthening Language with Sound

We're getting a better understanding of how language learning strategies improve sound processing in the brain. If we can predict the reading ability of a seven-year-old from the workings of her sound mind when she is a toddler, we can take action so negative predictions don't come to pass. The assistive listening systems used at the Hyde Park Day Schools are one of these strategies. The wearable word-counting technology from Providence is another. The auditory training games developed by Merzenich and Tallal and the toys that Benasich is developing for babies provide additional productive avenues. As we understand more about the sound-language connection, we can develop better ways to help children develop language skills.

Audio listening technologies are developing beautifully. I hope to see them become mainstream, not just in esoteric places like the Hyde Park Day School. One of my students has a language disorder and wears an assistive listening device that receives signals from a microphone I wear like a necklace while teaching. After class one day, I asked if we could trade, and was impressed by how clearly I could hear her voice from the other side of the lecture hall. I can imagine this technology could help *anyone* in a noisy place. All of us can benefit from stronger language skills.

As someone who thinks a lot about sound, I wonder what impact new ways of experiencing sound might be having on our listening brains. I've already noted that most days end for me with my husband reading to me. What I did not mention is that I also listen to audiobooks. What effect might this have on my sound mind, on how I read and speak and think? When it comes to comprehension and retention, it appears that listening to a text is similar to reading it.[67] In some cases, hearing might be better. I find that Shakespeare, with his archaic turns of phrase, is more understandable when heard than read. The addition of sarcasm, humor, or other cues in the voice of the actor also can help with a holistic interpretation of the heard passage. Reading aloud also improves your memory of what you have read.[68] I like to think we're more naturally inclined to understand and remember language through sound than print because our hearing evolved hundreds of thousands of years before we began to read and write.

Audiobooks expand the circumstances under which we have the opportunity to read. I use fitted earbuds that deliver sound and simultaneously block out the background noise while cooking (sizzling onions . . .), working out, on trains. I look forward to understanding the biological underpinnings of listening versus reading the same text and how this varies in individuals. I would like to know what listening to audiobooks is doing to the evolution of the sound mind.

8

Music and Language:
A Partnership

Musical training is a more potent instrument than any other.

—Plato

Music is a language.

"Who's got the voice?" my husband asks as he steps into the hall-way where I am talking to two men who have come to remove an old couch from our home. As each man speaks, my husband turns to one of them and asks, "Have you considered doing voice-overs, or using your voice professionally?" I had not picked up anything special about the voice, but it turned out the man indeed was a voice actor. Living with a musician is a constant reminder of how much sonic material many of us just aren't aware of. The two of us can be walking down the street together and hear a motorcycle. I just think "motorcycle" while my husband *hears* the make and model. The point is, making music hones the sound mind to non-musical sounds as well.

Music and Language

Music, for all its power to connect us, is not a good way to get certain information across. It is difficult to plink out directions to the train station on a piano or hum the score of last night's basketball game. Nevertheless, the relationship between these two modes of sound making is not accidental. Musicians enjoy distinct advantages in processing *speech* sounds, thereby strengthening communication through language.[1] The question is why.

The idea that music can influence language was formalized by Ani Patel in his acronymic OPERA hypothesis.[2] The O in OPERA stands for the overlap in brain networks responsible for music and speech. The P stands for the precision that music requires. We understand speech with foreign accents and bad phone connections, but a small distortion in timing or pitch or harmony can destroy musicality. Consequently, the enhanced precision that music demands positions the music maker well for making sense of other sounds. The E stands for emotion. Music engages reward centers that drive how we feel about sound. The R is for repetition—the honing of neural circuitry brought about by making sound-to-meaning connections again and again as we practice and play music. Finally, the A stands for attention, and we learn best what we pay attention to most. For all these reasons, it is fitting that an individual who spends a good portion of her day practicing music would hone her sound mind in a way that would aid the development of language skills.

Language affects music, too. Spoken English and French have different prevailing rhythmic patterns. (English is accent stressed, French is duration stressed.) The English Elgar and the French Debussy hew closely to their spoken language in terms of rhythmic patterns. The language of the composer leaves a mark on his music.

Language and music contain small units (phonemes; notes) forming longer phrases (words, sentences; musical phrases, songs) that convey information. In both, the way the small units are combined to form the longer units is governed by rules of syntax and semantics. Much as children without formal training learn to understand and create speech, we also, without training, learn to remember and reproduce music, to dance and tap along to music, and to feel the emotions it brings about. We can detect wrong notes and recognize violations in conventions of musical syntax with the same ease we recognize violations in the conventions of language. By studying music, we bolster these skills. Making music requires executing the intended notes at the correct times and developing the sound mind to distinguish correct and incorrect execution.

Similar sound ingredients underlie speech and music. Speech sounds are characterized by frequency (e.g., the difference between "ee" and "oo"), timing (e.g., the difference between "bill" and "pill"), or by the interplay between timing and frequency (e.g., the difference between "ball" and "gall"). The knowledge of the sounds of language, phonological awareness, is foundational to learning to read. A way to test knowledge of the sounds of language can go something like this: "Say the word 'please' without the 'l' sound." The ability to perform this task and other language sound-manipulation tasks is stronger in child musicians than age-matched nonmusicians, and relates strongly to the ability to discriminate melodies.[3]

In returning to the mixing board and sound ingredients, when we probe the musician's brain with the "mighty da," we find an enhancement of the sound ingredients crucial for language (see figure 8.1). Harmonics, one of these ingredients, helps to distinguish the sound of two instruments playing the same note *and* helps to distinguish speech syllables. The other key ingredients of timing and the rapid sweeps of frequency (FM sweeps) that occur as consonants transition into vowels, and vice versa, strengthen the musician's ability to make sense of the sounds in language.

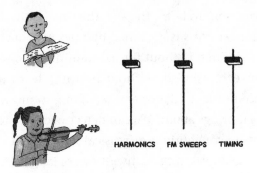

Figure 8.1
Language and music signatures coincide.

Reading and the Musician Brain

That music strengthens the processing of sound ingredients gives us some insight into why musician children outperform their peers on language skills. And moreover, music making can be critical to enhancing literacy. Playing music and reading both require forming *sound-to-meaning connections*. Before we become fluent, automatic readers, we spend a lot of time "sounding it out." We learn what sound a "T" makes, and an "R." And double E is easy. Putting it all together, we get "tree." Along the way, we learn which letter combinations make sense and which do not. We learn patterns and tricks like the "gh" in a "-ght" ending can be ignored, soundwise, as in "fight" and "caught." We learn, although perhaps only implicitly, that the choice between "im-" and "in-" depends on the following consonant: you might find a challenging book impressive and inscrutable, but never inpressive and imscrutable. There are analogous rules in music. Musicians learn to map musical notation to pitch and timing. The height (not pronounced heig-hit) of a note on the staff maps to how high or low its pitch is. The timing of a rest depends on whether a black rectangle sits atop a staff line or hangs down from it, much as we learn that a half circle to the left of a vertical line signifies a "d" and not a "b." Likewise, with experience, musicians learn that certain chord progressions and harmonic relationships, like the word "inpressive," simply do not "work."

In addition to the orthographic (i.e., sound/letter, sound/note) analogy between reading speech and reading music, there is the rhythm of speech. Every Martin Luther King Jr. Day, my husband and I listen to the "I Have a Dream" speech. Now, if I were reciting those words you'd be fidgeting, looking at your watch . . . This is because a huge part of the impact of that speech rests in King's rhythm. The rhythms involved in music making—and the resultant proficiency in rhythmic ability[4]—are key to language and to reading.[5] Children evaluated before and after music lessons, or rhythm-based training, show improvements in phonological awareness,[6] reading,[7] and neural processing of speech sounds.[8] The most challenging sound distinctions in language are the time-based ones involving consonant pairs like ba/ga or ba/pa. Thus, a musician's brain (attuned to music) makes a difference in how a child develops language and reads a book.

Auditory Scene Analysis: Hearing Speech in Noise and the Musician Brain

We live in a noisy world, and the number of circumstances where we have to struggle to understand each other—trains, airplanes, restaurants, classrooms, playgrounds—probably exceeds the time we spend in quiet. Our brains are pretty good at pulling the relevant sounds out of the irrelevant. This skill falls under the broader category of "auditory scene analysis," which is the way the sound mind organizes a soundscape into meaningful parts. By grouping the sounds of your conversational partner into an integrated object, you can home in on his voice against the backdrop of other conversations swirling around you at a party. This ability is better in some people than others, though, and as a whole, musicians are especially good at this challenge.[9]

This listening advantage is not limited to highly trained musicians. There are similar benefits for beginners. Brainvolts had the opportunity to investigate listening to speech in noise in grammar school children who were just starting out as musicians. We assessed their listening in noise ability before they started their music training, then after a year, and then again a year later. There was perhaps the merest hint of improvement after one year. But after two full years of regular music making, children demonstrated marked increases in the amount of background noise they could tolerate and still repeat the sentences back accurately.[10]

When listening conditions are good, the differences between musicians and nonmusicians are not as obvious. The brain "lights up" equivalently when musicians and nonmusicians listen to speech in favorable conditions; only with added noise does the musician's edge appear (figure 8.2).[11] A similar pattern is seen in the brain's physiological response to sound.[12] The nonmusician brain in the unfavorable listening condition shows a diminished response in both the brain image and the physiological waveform.

Why are musicians' sound minds so good at hearing *speech* in noise? The OPERA hypothesis gives us some clues, and I would like

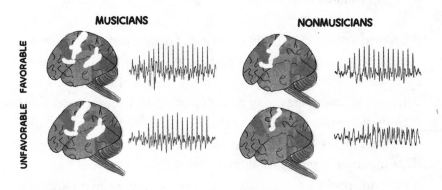

Figure 8.2
Musicians exhibit stronger responses to sound (the white in the brain images) in unfavorable listening conditions (bottom row). This is mirrored in their physiological waveforms.

to suggest adding rhythm and working memory—vital skills for musicians—as two more key components. (I'll need to work on the acronym; OPERRAW feels undercooked.)

The rhythm of speech allows us to fill in the gaps in noise. When noise obscures speech, the underlying rhythm helps us predict the words we cannot make out. Accordingly, drummers, it seems, are especially good at hearing speech in noise.[13]

If you're good at working memory—essential for following a conversation—you are able to listen in noise better, whether you are a musician or not.[14] Playing music is a good way to strengthen memory skills.[15] Making sense of sound places a strong demand on our ability to think. Joe Saxophone, with better working memory, has more processing power to throw at any task. Musicians' superior ability to track changes in pitch contours[16] and sound patterns[17] contributes to their ability to hear the longest and most semantically complex sentences in noise.

Although there is not complete agreement on this,[18] in my view, evidence converges that musicians—whether due to a combination of enhanced sound processing in the brain, rhythm facility, working memory prowess, or something we have not yet discovered—are simply able to hone their sound minds to efficiently analyze auditory scenes.

Within musicians, both the extent of practice and the age at which they started playing correspond with listening-in-noise ability. These relationships show that benefits continue to accrue with experience.

Often, despite the best of intentions, we stop making music as we get older. Still, the positive outcomes of making music last, to at least some extent, *even if you do not continue to play music*. Playing music is a good investment that pays off in young adulthood[19] and even decades later.[20] Once the brain has learned to make strong connections between sound and meaning, the brain continues to reinforce this skill automatically.

Neuroeducation

Teachers resoundingly tell me that children who play music do better in school. They tell me they see this every day and are frustrated others cannot appreciate what is so obvious to them. They ask me, "What is going on in the brain?" About ten years ago, I was approached by a force of nature, Margaret Martin, the founder of the Harmony Project in Los Angeles, a nonprofit program devoted to getting instruments into the hands of underserved children and assuring they have the very best music instruction. With a doctorate in public health, Margaret documented student outcomes meticulously. She knew firsthand that music *just worked* for academic advancement. "Nina, I know music can keep at-risk youth from dropping out of school," she told me, "often making them the first in their families to go to college. With your help, we can get to the bottom of *how* it works, so we can become more successful in getting the word out." And so our collaboration was born.

Meanwhile, I had almost the same conversation with someone closer to home, Kate Johnston, a music director in the Chicago Public School system. She taught at a school that featured music instruction as an integral part of the school day, alongside English, history, and math. Thus, at more or less the same time, Brainvolts embarked on two major, logistically challenging, longitudinal neuroeducational projects on the influence of musical experience on the sound mind.*

Music in Natural Settings

We were keen to undertake these studies because both offered an uncommon opportunity to look into the effects of musical experience on the nervous system in *natural* settings. By natural I mean

---

*Neuroeducation uses neuroscience to understand how learning takes place in the brain with the aim of improving teaching methods and academic achievement.

long-standing, successful music programs, not artificial programs designed by scientists. This was a chance to learn about the biological basis for the interplay of music, learning, and educational achievement in the real world through the lens of the sound mind.*

The children in the Harmony Project were young—second graders—and they were new to music instruction. The project was led by then–graduate student Dana Strait, who led her team of four to conduct their research, no joke, in a storage closet. Over the course of three years, they would clear out the mops and vacuums, packing boxes, half-dead computer monitors, and broken musical instruments to set up shop. In this far from acoustically and electrically shielded space, they would test students in three-hour sessions, on their listening-in-noise ability, reading and cognitive skills, and their brains' responses to sound.

The Chicago high-school project was closer to home, but the protocol was four times larger in scope. Chicago Public School students participated in the research as entering ninth graders until their graduation from high school. For most of them, that first year of high school was their first experience with music instruction. We did much of the testing in Brainvolts' home base. Led by Jennifer Krizman,† we also organized periodic "testing fairs" at the schools involved in the study, all-hands-on-deck affairs with a dozen Brainvolts students and staff caravanning to Chicago neighborhoods with computers, stacks of testing materials, neurophysiological equipment, and enough food to keep everyone functional for an all-day session of nonstop data collection. We kept this up *for five* years, testing upwards of 200 unique participants annually, over multiple test sessions.

---

*Many constraints make it difficult for scientists to do their work in educational settings. Our opportunities were uncommon because the educational programs themselves invited *us* to come to *them.*

†Jen got to know her high-school student participants. She called and texted them so much—it can be hard to get a teenager to follow through—she committed many of their phone numbers to memory. She attended all their graduations and wrote many letters of recommendation.

Somehow, for both projects (did I mention they were being conducted at *the same time*?) we managed to successfully move teams and
equipment hither and yon, and ensured nothing crucial changed from
year to year. All the while, Jen and Dana needed to keep all the other
personalities involved (teachers, parents, administrators, janitors . . .)
happy and coming back for session after session, year after year.

Are Musicians Made or Born?

One of the biggest criticisms lodged against the "musician advantage" is that of causality. Correlation does not mean causation. Jody,
with twenty years of playing piano, has more white matter than Pete,
who has never picked up an instrument. Does that mean Jody's brain
developed more white matter due to playing music? Or was she born
that way? It's possible that something about the way white matter
works in the brain steered her to an interest in music that landed her
in front of the piano in the first place. Did four-year-old Fred, with
his super large right motor cortex, feel compelled to drag his parents
to the luthier from some kind of biological imperative?

It would be impossible to argue there is *no* "nature" aspect involved
in who is drawn to music. People's brains and bodies can predispose
them to becoming musicians. But as my husband has observed from
decades of teaching music, it's the ones who *want* to play who make the
most progress. Learning what we care about shapes the sound mind.
The focus of our work, then, is on the "nurture" side of the equation
because nurture is where we have the power to do something.

The causality, or nature vs. nurture, question can mostly be put
to bed by looking at longitudinal investigations because it is possible to compare participants to *themselves*. Longitudinal studies
provide strong evidence that the "nurture" of music education can
reshape the sound mind no matter what the starting point. Comparison groups typically engage in another healthy activity for the same
amount of time as the musicians (figure 8.3).

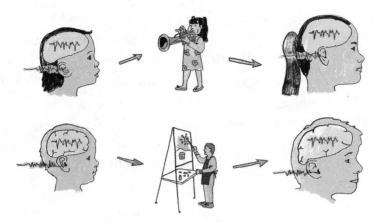

Figure 8.3
A neuroeducational longitudinal design.

What We Learned

Strengthened sound processing in the sound mind can contribute to the improved academic and listening skills seen in musician children.[21] In both the Los Angeles grade-schoolers and the Chicago teens, *enhancement in processing certain sound ingredients* in the brain was found in the music-making children only. These are the very sound ingredients needed for reading and language development (figure 8.1).[22] The brain became better tuned to the harmonics we use to identify speech sounds and to better track timing cues and the FM sweeps that mark the transition from consonants to vowels and back again. Moreover, these effects occurred even when music training began later in childhood, in our high school kids, documenting the flexibility of the brain for auditory learning.

Brainvolts is not the first or only group to do longitudinal research into music's impact on the brain and on language skills. Mirielle Besson's group in France found processing timing and duration cues in speech (but not pitch) is enhanced after a year of music training in eight-to-ten-year-old children.[23] An increase in sound processing in the brain corresponds with increases in verbal intelligence, reading, and cognitive skills not found in a control group who had a similar duration of art training.[24] Others have noted gains in attention and memory,[25] auditory processing,[26] second language learning,[27] vocabulary,[28] responsibility and discipline,[29] and the ability to block out irrelevant sounds.[30] Longitudinal investigations have reinforced the accelerated brain maturation we observed,[31] and John Iversen (of Snowball the cockatoo fame) is spearheading the SIMPHONY project in San Diego to investigate the development of the brain in child musicians.

Music Can Offset the Neural Signature of Poverty

Poverty puts people at risk for many health hazards, including impediments to the sound mind.[32] In our Chicago and Los Angeles music projects, the children lived in low-income neighborhoods, attending schools where more than 85 percent of the families qualify for subsidized lunch.

HARMONICS   FM SWEEPS

Figure 8.4
The brain signature of linguistic deprivation (top) can be offset by music (bottom).

Study participants showed diminished responses to key sound *ingredients* in speech. These included harmonics, slower neural timing to consonant-vowel transitions (FM sweeps), and diminished neural stability (consistency).[33] Compounding the weaker processing of speech is excessive neural noise, which can be thought of as brain static. The neural signature of poverty is illustrated in figure 8.4, where the faders on our mixing board are diminished for sound ingredients while the neural noise is turned up. The diminished signal and the excessive noise are liabilities that impede sound processing as illustrated by figure 8.5.

The musician makes the sound clearer by processing key sound ingredients more effectively, thereby turning up the sound. Thus, making music partially offsets the poverty signature by strengthening the brain's response to harmonics and crucial timing cues, although it does not boost consistency. Other strategies can offset the neural signature of poverty. Being bilingual can turn up sound in the brain, and athletic participation can turn down the noise. An examination of the sound mind's processing of *sound ingredients* can provide insight into the different and complementary mechanisms invoked by musicians, bilinguals, and athletes.

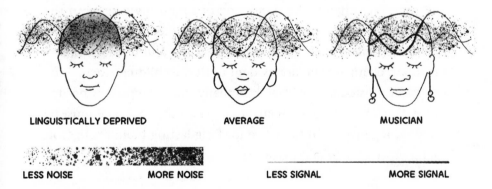

LINGUISTICALLY DEPRIVED   AVERAGE   MUSICIAN

LESS NOISE   MORE NOISE   LESS SIGNAL   MORE SIGNAL

Figure 8.5
Compared to a typical brain (center), poverty (left) turns up the noise in the brain and turns down the signal, while playing music turns up the signal (right).

Closing the achievement gap   Children from low-income neigh-borhoods often perform poorly on reading and other academic skills compared to their more privileged peers.[34] This achievement gap only widens as the children get older.[35] In Los Angeles, second graders from low-income families show a decline in reading scores, which is, regrettably, a typical progression. In contrast, the children in Harmony Project maintained their reading skills.[36]

You're Not Going to Get Physically Fit *Watching* Sports

Listening to music has a place in relaxation, stress relief, and mood regulation.[37] It can also provide temporary benefits in attention, memory, motor synchronization, and reasoning skills,[38] likely due to the increase in dopamine production brought about by listening to enjoyable music[39]—being in a good mood confers a bump in think-ing ability.[40] Moreover, music listening can aid in the treatment of neurological conditions such as dementia and Parkinson's disease, and in recovery from stroke.[41] However, despite popular notions about the value of exposing babies to classical music in the crib or even in utero, so far there is a distinct *lack* of evidence that merely *listening* to music has much lasting impact on the sound mind.

As Harmony Project made clear, there must be *active engagement* with music to change the processing of sound ingredients in the brain. The project began with music fundamentals training that included careful and directed music listening and not much play-ing. Brain changes were not evident until the children reached the hands-on music-making stage.[42] Actually making music is needed to create changes in our default response to sound. There must be training, repetition, and practice for long-lasting brain changes in sound processing to take place.

It Takes Time to Change the Brain

Reviewing résumés and graduate school applications is part of my job. Increasingly, people report many experiences, spending a little time on more and more endeavors. Many have spent what seems like five minutes in Ecuador, another five minutes being a camp counselor, another five making pottery. Yet, in my experience, the strongest students are the ones who have stuck with one or two activities for a long time.

In both longitudinal studies in Chicago and in Los Angeles, changes in sound processing in the brain were not evident after one year of music instruction. *Fundamental changes in how the sound mind processes sound ingredients essential to language were observed only after two years of music making.*[43] The implication is that the impact of music education on the brain cannot be achieved quickly or by flitting from one activity to another, no matter how beneficial those activities may be. While this slow pace of change may seem like a negative, there's a plus side to it—imagine how confusing it would be if our brains changed in fundamental ways from moment to moment. Who we are from a biological perspective takes long-standing and persistent engagement.

Embracing Music

If I had five minutes to tell teachers, parents, health-care workers, and anyone who would like to listen to me about the biological evidence that supports the benefits of music education, I'd say:

- The sound mind is vast. It engages how we think, feel, and move. Music is uncommonly good at engaging the entire sound mind. Through this engagement, making music shapes brain networks to strengthen sound processing.
- The skills and brain activity improved by making music are many of the same that are needed for language and reading.
  - Making music can boost academic achievement.
  - Making music can help bridge the academic achievement gap between the rich and poor.

- The brain signature of making music is evident regardless of the
  ○ instrument played (yes, voice counts);
  ○ genre of music played;
  ○ instruction type (group or individual);
  ○ instruction modality (in-classroom or private lessons); or
  ○ instructor (public-school teacher or musician without formal teaching credentials).
- *Active* music making changes sound processing in the brain. Passive listening is not enough.
- The effects of making music persist beyond childhood.
  ○ Making music strengthens sound processing in the brain long after music making has stopped, even into old age.
- Music making is not a quick fix. It takes time and persistence to change the brain.
- Music does an uncommonly good job fostering a sense of community.
  ○ It creates social cohesion and singleness of purpose in groups, by drawing individuals into a greater whole and making them emotionally predisposed to work together.
  ○ It is a universal language with traditions in all cultures. This is because the sounds of music carry with them engagement of our feeling and cognitive systems.
  ○ Moving in sync predisposes us to cooperate.
- Beyond scientific evidence there is economics to consider.* Music education can help keep kids out of trouble, at a fraction of the cost of medication and incarceration.[44]

---

*It costs $35,000 per year in the US to keep someone in jail. The total cost of imprisonment, including court costs, policing, parole, and bail, is estimated at over $180 billion annually. If you add in societal costs, the financial burden of incarceration is estimated at a staggering $1 trillion annually. Medication to address attention problems in the US costs $20.6 billion per year.

There is an *intangible argument for music education*: some of the deepest benefits of music education are challenging to quantify.[45] Music supports child development in its most holistic sense— the lasting friendships, the focus and discipline that comes from years of regular practice, the social engagement from playing in an ensemble, the confidence that develops from performing on stage. Music brings a new dimension of education to children not found in any other school subject. The movement of playing an instrument is a nonverbal form of thinking and knowing—a means toward a higher and fuller consciousness, a self-knowledge of feeling, and a way to develop aesthetic sensitivity.[46] As educator Bennet Reimer put it, "Music belongs to basic education because musical experiences are necessary for all people if their essential humanness is to be realized."[47] These intangibles are real, but unlike cognitive and language skills, they cannot be easily measured. It is unlikely there will be a clinical trial to determine whether music is effective at moving the needle on these less palpable benefits. In fact, every layer of control added to experiments about music training often obscures the intangibles that make music music.

Despite being a scientist, I believe the tools of scientific inquiry are not appropriate to answer every question. The unmeasurable benefits of music making are no less real and no less important than the measurable ones. And, I have every expectation these intangibles contribute to the measurements we *can* make.

Music Education From the Point of View of the Sound Mind

What would ideal music education look like if we had all the resources to make it a fundamental part of education? Music education, in its simplest form, does not require fancy instruments or equipment. The first instrument a child has is their voice—indeed the first instrument humans had was their singing voice, predating the first musical instruments by millenia.[48] Rhythm requires nothing more than your hands or some pots and pans and wooden spoons. It's never too early to start making music.

Isabelle Peretz who has spent the last three decades studying the neurocognition of music—from musical prodigies to people who cannot carry a tune—maintains that everyone is musical. At the two extremes of a normal distribution, 2.5 percent of the population can be considered musically gifted, 2.5 percent of the population would be amusic. Said Isabelle, "It is important to note that the vast majority can reach a professional level if they invest enough practice hours."[49]

Teaching music is a means of enculturation, of creating a sense of community and belonging. An excellent curriculum is an excellent teacher.[50] In my view, we must have *an educational system that prizes the excellent music teacher*.

We routinely ask our research participants the extent to which they play music "by ear" or through sight reading.* I have been surprised to learn that almost everyone falls into one camp or the other. Why not teach *both* approaches together? Children love to imitate. Show them how a piece is played and let them copy you. Then show them how it's written to connect the two. This is what I call being musically bilingual. Being able to play by imitating, reading notes, and improvising enlarges the range and contexts for making music. While either musical approach on its own can change the sound mind for the better, the musically bilingual person seems to have an especially well-tuned hearing brain.[51] Reading notes provides a common language. Indeed, reading music and reading words are operations that share similar but not wholly overlapping brain resources and practice in one strengthens the other.[52] I'm grateful for my

---

*I don't much care for the expression "by ear." Let's call it learning by listening and imitating—"imitating" will suffice. It is how we learn most things, e.g., the language we speak.

piano teacher who helps me work out chords and harmonies in rock and jazz, guides me in improvisation, and teaches me Beethoven.

Surprisingly, there do not seem to be many mechanisms to encourage moving from academe or performance to music education or medicine. Steady work in education or medicine could help fuel the performing musicians' life. There are barriers that separate musicians, academicians, music therapists, and clinicians. These experts all bring something to the treatment of children with language disorders and adults recovering from stroke, for example. A move toward earlier and stronger music education—teaching by imitation, note reading, and improvisation—and incorporating a range of musical styles, should help bring the various groups together. I suppose my view on music pedagogy aligns with my scientific preference for operating at the meeting point of disciplinary borders.

Making music changes the sound mind for the better. Speaking as a scientist, music should be taken seriously in education and medicine.

9

The Bilingual Brain

One egg is un oeuf, but is one language enough?

If I could choose a superpower, it would be to have the ability to speak any language.

In his book *Born a Crime*, Trevor Noah recounts how he was able to transcend the color cliques in his high school with language. Race tensions in South Africa meant a polarized linguistic environment, with whites speaking Afrikaans while blacks mostly stuck to their tribal languages for all but official purposes. Noah, who is mixed race, speaks both Afrikaans and Xhosa, and this gave him an in with both whites and blacks in his school. His language got him accepted as one of the guys among classmates of any skin tone and made him one of the few able to move in both white and black social spheres. I would love to speak the language of anyone I meet so I could connect with them on the deep level achievable only through a shared language. This sense of *belonging* stems, at least in part, from shared sound mind circuitry tuned to the same sounds.

Worldwide, more than half of us speak more than one language.[1] We see a different story in the US, where only one in five people speak more than one language.[2] In what ways might a bilingual brain differ from a monolingual brain? If by speaking a second language we boost our vocabulary, add to our grammatical playbooks, and augment our repertoire of language sounds and perspectives, do we give up something else? These questions have been around for a long time.

Whether for reasons of perceived negative economic impact or fear for safety, demonizing the "other" has been a hallmark of human life since earliest recorded history. The word "barbarian" arose from the "bar-bar-bar" sounds the Greeks perceived outsiders speaking, with the subtext that outsiders were not intelligent enough to speak a proper language. As recently as the middle of the twentieth century, a prevailing scientific opinion in the US was that foreign-language speakers, even those who spoke English well, had inferior mental abilities to native English speakers.[3] "There can be no doubt that the child reared in a bilingual environment is handicapped in his language growth," stated a child-psychology textbook in 1952.[4] Much of the anti-bilingual bias was due to negative attitudes toward the increase in immigration from southern and eastern Europe. The congressional Dillingham Commission, convened in 1907, concluded that immigration from these regions was a serious threat to American society. Tests of English vocabulary and English-centric knowledge were pulled together and, based on the results, conclusions were drawn that the European immigrants to Ellis Island at that time were "feeble-minded" compared to the settled and assimilated Anglo-Saxon and Nordic immigrants who had arrived forty years earlier.[5] The commission's conclusion led to literacy tests and quotas that decreased immigration rates across the board and all but stopped immigration from Asia for several years.

In the years since, there has been a softening in stance. It is generally agreed bilinguals have an edge in some things but might be at a disadvantage in others, if you control for things like length of time living in the US, a consideration that eluded the Dillingham commissioners a century ago.

Let's view bilingualism through the lens of the sound mind and see what it can tell us about this language superpower.

Language-Specific Tuning of the Sound Mind

Any person can speak any language—the anatomy involved is universal. However, it can be difficult for an adult to adapt to the sounds of a new language. Almost any pair of languages will have some mutually incompatible sounds. Take the timing ingredients in sound. The time the vocal cords start vibrating, relative to the opening of the lips (called voice onset time) distinguishes "bill" from "pill." If you start the voicing right away, you get a "b"; if you wait a short amount of time, you get a "p." Some languages additionally feature *pre*-voicing. You actually begin the vibrations *before* you open the lips. Pre-voicing is more or less undetectable to an English speaker; it still sounds like a "b." However, in Hindi, to give one example, a pre-voiced consonant is a valid class of sound easily distinguished from the others.[6] In the English language, this distinction is of no use, so the sound mind of the English speaker does not waste energy making the distinction.[7]

The principle of sorting sounds of language into perceptual categories is called (surprise) categorical perception.[8] In English, we can turn "bill" into "pill" by adding 50 ms of silence. What would happen if we added 25 ms? Would we hear some kind of bill/pill hybrid? This has been studied exhaustively and the answer is essentially no. If we manipulate the length of the silent gap from say 0 to 50 ms in 5 ms steps, we will usually see an all-or-nothing change in perception from "b" to "p" at somewhere around 30 ms. Everything from 0 to 25 is "b"; everything from 30 to 50 is "p" (figure 9.1). No voice onset time produces an ambiguous bill/pill hybrid. A typical English speaker on a "same or different" task would have no difficulty choosing "different" when listening to a 20 ms and a 30 ms

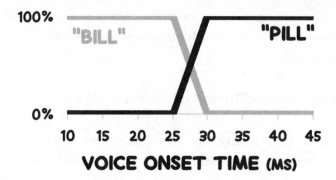

Figure 9.1
Categorical perception. With increasing voice onset time, our imaginary subject heard "bill" (gray line) 100 percent of the time until the timing cue hit 30 ms. Then "bill" perception dropped to zero, and he unambiguously heard "pill" (black line) 100 percent of the time.

pair because it crosses the categorical b/p boundary. But that same 10 ms difference would register as "same" if the pairs were 30 and 40 ms. Both sounds are in the "pill" bin and, because of our experience with language, they are perceptually difficult to distinguish.

That is why pre-voiced consonants in Hindi are difficult for native English speakers. We have not developed a category to slot them into. Despite initially falling into an English speaker's "voiced" category, we can learn the voiced vs. pre-voiced distinction with practice. Brainvolts alumna Kelly Tremblay trained English speakers to hear the pre-voiced distinction Hindi speakers can natively discern. Given enough training, they could hear the distinction, and their auditory-brain responses changed accordingly.[9]

Sound *perception* can transfer to sound *production*. Training a Japanese speaker to hear the English "r"/"l" distinction benefits their production of those sounds—the highly interconnected sound mind at work.[10] I personally experience this auditory-motor, hearing-talking interaction when I travel between the US and Italy. When I first arrive in Italy, my mouth feels like it is full of marbles when

I speak Italian. But after a few days of immersion, my spoken Italian begins to flow smoothly. I experience the same fluency lag when I get back to the US.

The brain produces an MMN (mismatch negativity) to a change in an otherwise regular sequence of sounds. The more acoustically different the sound pair is, the bigger the MMN. Recall that the relative sizes of the bumps in a speech sound's spectrum are the acoustic ingredients that determine which speech sound was spoken. The Estonian language (figure 9.2, top) has four vowels, /o/, /õ/, /ö/, and /e/, with bumps in the spectrum centered at approximately 850, 1300, 1500, and 2000 Hz, respectively. The sound mind of an Estonian will distinguish these vowels in a systematic way. More acoustically disparate pairs (e.g., /e/ v. /o/) elicit a larger MMN than closer pairs (e.g., /e/ vs. /ö/).

However, language experience subverts this principle. Estonians share many similarities in their language with their neighbors, the Finns. Both languages share /e/, /ö/, and /o/. But the Finnish language (figure 9.2, bottom) lacks the /õ/. Based on acoustic differences we should see MMNs increase in size as we compare /e/ to /ö/, /õ/, and /o/, respectively. And for the Estonians, that's exactly what we see. But this isn't the case for the Finns. The /õ/ response is not larger than for the /ö/; in fact, it is smaller. So it's not just about how acoustically different the sounds of the vowels are, but whether or not they have a home in your sound mind. The brain is more tuned to the sounds of your language than to sounds outside your language.[11] Similarly, voice pitch, which carries phonetic meaning in Mandarin but not in English, induces more robust MMNs in Chinese listeners than in American listeners.[12]

When does this language influence on auditory processing develop? To answer this question Risto Näätänen's group led a study looking at MMNs to those same vowel sounds in Finnish and Estonian infants. At six months of age, the brains of Finns and Estonians

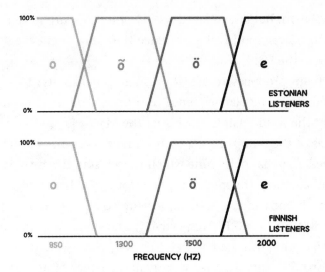

Figure 9.2

Four vowels, /o/, /õ/, /ö/, and /e/, have frequency bands centered at approximately 850, 1300, 1500, and 2000 Hz. A speech sound with a peak at 1300 Hz would be identified by an Estonian 100 percent of the time as /õ/. In contrast, the Finnish language (bottom) lacks an /õ/.

processed the /õ/ identically. But by one year of age, the Estonian vs. Finn adult pattern had emerged.[13] Nearly identical results have been found in other languages. For example, American and Japanese babies equally distinguish "r" and "l" at the age of six to eight months. However, by one year of age, the American babies are doing much better while the Japanese one-year-olds' ability to distinguish them is worse than at the starting point.[14] Language-specific sounds begin to form in the sound mind early in life.[15]

This goes a long way toward explaining why starting to learn a second language at a young age is best. Learning new sounds and developing categories to slot them into is easier for a younger brain. Someone who learns a second language at a young age will almost always have less of an accent than someone who starts as an adult, because they can nail the subtleties of the sounds of the language being learned before experience with their native language has solidified categories that are difficult to break out of.[16] Speaking another language when we are younger also means more time spent making sound-to-meaning connections that change our sound minds accordingly. As in musicians, the age of acquisition of the second language and how long it has been spoken are important considerations.

The Bilingual Brain Is Not the Sum of Two Monolingual Brains

If you are a bilingual having a conversation in one of your languages, do you fully switch off the other language? Some believe it is possible, but there is increasing evidence the other language is never fully switched off.[17] Both languages remain "available" to the bilingual speaker even if only one is being used at a given moment.

Imagine you are shown a grid of fifty or so pictures on a computer screen. These are all everyday objects, animals, etc., and you are tested on how quickly you can select the correct object when you hear a word spoken. Get ready, here comes the first word, "cah . . ." Already, before the word is even finished, your eyes are darting around the screen, narrowing down your choices to "coffin" and "coffee" and "cobweb." The initial "cah" sound has primed your brain for the limited set of objects that begin with that sound, even before the word is completed. But if you are a bilingual English-Spanish speaker, the "cah" sound will additionally activate your store of Spanish vocabulary. You will not be able to immediately rule out the pictures of the horse (*caballo*), the truck (*camión*), the puppy (*cachorro*), or the box (*caja*) that never for an instant distracted the monolingual English speaker. So while you can still perform the task without any errors, it is a bit more effortful; you probably are a bit slower at it because your first pass at narrowing down the field left you with seven contenders instead of just three. A design similar to this, using eye-tracking, confirms bilingual speakers indeed take a longer look at objects with spelling/sound similarities in the language that is not being tested.[18]

We see cross-language interference biologically. Remember that the brain is especially good at detecting a change in a predicted sequence. There is a brain response called the N400 (a negative-going brain wave that occurs 400 ms after sound onset) that signals *semantic* incongruity, in contrast to the MMN, which signals acoustic incongruity. If you listen to the sentence "The airplane landed at the airport," there will be no N400; the sentence has no semantic violation. However, "the airplane landed at the grapefruit" will trigger an N400 because it violates semantic expectation. In a clever study, researchers capitalized on this neural response to look at cross-language interference. They quizzed Chinese-English bilinguals on whether English word pairs were semantically related (wife-husband) or unrelated (train-ham). But they chose the word pairs very carefully. In some cases, the Chinese equivalents were logographically and phonologically similar. The Chinese translations of train (火车) and ham (火腿) both begin with the same character and same pronunciation, something like "huo." This Chinese-language similarity affected their N400 responses, suggesting they spontaneously accessed the words' Chinese translations. Word pairs that were incongruent in English, but had similarities in their Chinese counterparts, had smaller N400 responses than word pairs that were equally incongruent in both languages, like apple-table (苹果—桌子).[19] Knowledge of Chinese did not interfere with the sound mind in its treatment of these English pairs.

Thus, bilinguals do not fully "turn off" one language in a situation where the other language alone is called for. But let's consider the implications. Just because you're slower to react to words when there are added lexical possibilities doesn't mean it's a "bad" thing—unless you're hell-bent on being the fastest in a reaction time experiment. Those other linguistic possibilities likely provide richer ground for thinking, conjuring memories, and other associations that go into our sound-to-meaning connections. A bilingual brain, then, is not the sum of two monolingual brains. The two languages comprising a bilingual brain interact with one another in ways that can be both beneficial and problematic. Speaking more than one language influences how sound makes us feel, think, and move.

The Downside: What Do Bilinguals Give Up?

A bilingual generally has a smaller vocabulary in each language than a monolingual,[20] a consequence of spending less time speaking any one language. This can be problematic because a reduced vocabulary might inaccurately be construed as a language disorder. There are related difficulties in word retrieval. Quickly and fluently coming up with the desired word is more challenging for bilinguals,[21] presumably because of interference from the other language.[22]

Bilinguals also seem to have a more difficult time understanding speech in background noise than monolinguals.[23] Consider the case where the background noise is the sound of other voices. You are a Spanish-English bilingual having dinner in a noisy restaurant with your English-speaking friend. You are at a double disadvantage— *imperfect signal* and *imperfect knowledge*.[24] You probably have a smaller vocabulary in English. Yet your vocabulary *overall*—across both languages—is greater.[25] So you and your friend are talking about horror movies and she mentions she has seen *Misery*, and you are wondering why she is suddenly talking about not having access to her iPhone assistant ("seen *Misery*," "*sin mi Siri*"—"without my Siri"). OK, that example was cringey and contrived, but you get the idea. Reduced exposure to the language, in this case English, gives the bilingual less familiarity with linguistic cues that help fill in the gaps when listening in noise. Less knowledge of the language, coupled with activation of a greater number of linguistic competitors, leads to difficulty hearing speech in noise (figure 9.3).

Interestingly, however, when language is not involved, bilinguals are *more* proficient at listening in noise. When English-Spanish bilingual teens were asked to perform a *nonlinguistic* listening in noise task—namely, detecting *tones* masked by noise—they outperformed their monolingual peers (figure 9.3).[26] This suggests bilinguals' experience with a richer repertoire of the sounds of language can bolster auditory processing in noise as long as it is not in the realm of language, where cross-language interference can undermine sound processing.

The Upside: What Do Bilinguals Gain?

If I am bilingual, I can speak to more people than if I am monolingual (cue superhero theme music). This is an obvious advantage and one that is motivation enough for many to learn a second language. But there is reason to believe speaking two languages confers other benefits. Why? Learning a second language calls on many of the same elements—exercising attention and memory, expertise

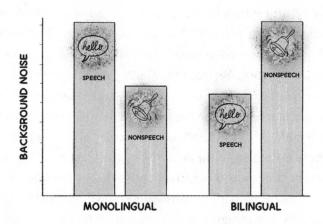

Figure 9.3
Bilinguals are better able to hear nonspeech sounds in background noise than monolinguals, although their ability to hear speech in noise is worse. Y-axis is level of background noise that can be tolerated.

in sound processing, getting the neural circuits sparking—as music making. Like music making, there are collateral benefits to speaking a second language.

The sound mind operates in alliance with how we think, sense, move, and feel. Let's start our survey of bilingual advantages with thinking. Cognition includes attention, working memory, planning, organizational skills, flexibility in thinking, self-monitoring, and the ability to ignore irrelevent information. Speaking another language can enhance these abilities and help you *think* better. Studies have focused on many aspects of cognition in bilinguals with competing viewpoints for many of them,[27] but one that most frequently rises above the hubbub is attention.

Bilinguals excel at suppressing impulsivity, which is key to being able to avoid distractions and pay attention to what's important. This skill is called "inhibitory control." A favorite assessment to measure this skill is the dimensional change card sort task. Despite the intimidating mouthful of a name, the task is simple. There is a stack of cards with shapes of different colors. Your job is to sort them by shape. All the diamonds go in a pile, all the squares go in another pile . . . regardless of their color. Then you are told to do it again, but this time sort by color. Assemble all the blues, all the greens . . . ignoring the shape. In this task and a range of others that challenge inhibitory control, bilinguals outperform monolinguals. And bilingual children can perform the task at a younger age than their monolinguals peers.[28] This advantage makes sense when you consider that a bilingual must suppress the vocabulary and syntax of Language 1 when speaking or writing in Language 2 and vice versa.[29]

The bilingual sound mind also excels at navigating sound patterns. The skills required to discover patterns in artificial languages are heightened in bilingually raised toddlers[30] and adults,[31] suggesting that once you have learned a second language, learning additional languages becomes easier.[32]

The *auditory scaffolding hypothesis*[33] holds that experience with sound, especially language, is a foundation upon which cognition is built. Deaf children have problems with attention skills, even some that are explicitly *visual* in nature, lending support to this idea.[34] And as people grow older, speaking more than one language may bolster cognitive skills and stave off cognitive decline.[35]

Brainvolts looked for biological markers of bilingualism in the sound minds of children and adolescents. Jennifer Krizman (then a student, now a Northwestern University professor) led the bilingual work and lived and breathed this venture for half a decade. She reasoned that music lessons as a means to enrich the sound mind can be cost-prohibitive for many families, and this is especially true in the immigrant demographic. However, immigrants very often speak two languages. Jen wanted to find out whether speaking a second language could offer benefits to help counter the stigma in the US surrounding bilingualism. Being bilingual could provide an opportunity to strengthen the sound mind when more expensive methods are not an option. She wondered whether certain sound ingredients are processed distinctly by the bilingual brain.

The telltale signature of the bilingual brain is increased sensory processing of the *fundamental frequency*[36] and a highly stable or *consistent*[37] response to sound (figure 9.4). In speech, the fundamental

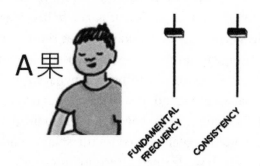

Figure 9.4
Bilingualism strengthens the consistency of sound processing and the response to the fundamental frequency (pitch cue).

frequency—voice pitch—is a strong language marker. Different languages have, on average, higher or lower pitches,[38] and speakers of two languages will nearly always speak one at a higher average pitch than the other,[39] supporting the importance of this sound ingredient for a bilingual. The fundamental frequency also helps us distinguish one "auditory object" (David's voice, the roar of traffic, Sara's voice) from another. Distinguishing one auditory object from another is more challenging than distinguishing visual objects. Determining with our eyes where one car stops and another car begins is pretty unambiguous unless there has been a ghastly crash. Distinguishing the sound of two cars, if achievable at all, likely comes down to the pitches of their engines, exhaust systems, and the distinctive sounds of tire treads on pavement. As we noted above in the case of a tone masked by noise, these sorts of auditory-object distinctions are easier for bilinguals. As for consistency, a well-tuned auditory brain will respond identically to a given sound each time it occurs; deviations from this fidelity would signify a lack of consistency. In brain activity originating from both subcortical (midbrain) and cortical auditory areas, bilinguals exhibit a more consistent response to repeated sounds. Both findings, a stronger response to the fundamental frequency and increased consistency of processing, are directly correlated with performance on measures of attention, inhibitory control, and language proficiency.

How does being bilingual affect the sound mind signature of poverty? Children raised in poverty can exhibit diminished processing of several sound ingredients, including harmonics, FM sweeps, and consistency. When we dug into our data set of Chicago and Los Angeles public school children with an eye toward second-language experience, the hallmark brain signature of poverty was simply less apparent in bilingual youth. In monolinguals, children from higher socioeconomic families had more consistent neural responses to sound than children from low-income families. That difference, however, was largely absent in bilinguals. The low-income bilinguals' consistency, in fact, matched that of the high-income

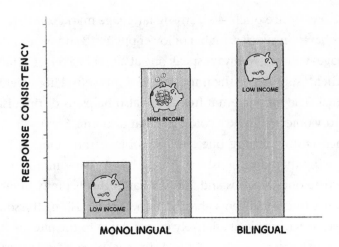

Figure 9.5

Bilingualism seems to have a protective effect on response consistency, regardless of socioeconomic status.

monolinguals (figure 9.5).[40] The edge seen in sound mind processing in bilinguals likely stems from the experience with a greater set of phonemes—language sounds—than monolinguals. More brain resources are exercised in processing a richer set of language sounds, leading to heightened sound mind processing. The low-income bilinguals also outperformed high-income monolinguals on cognitive tests (attention and inhibitory control), in agreement with others.[41] Thus, speaking another language can offset the neural and cognitive signatures of poverty—did someone say superpower? This advantage is driven by a more consistent response to sound.

Being bilingual confers cognitive and sensory benefits in the sound mind. What about moving and feeling?

When we speak, we *move*. I cannot be tethered to a podium when I give a talk. I was once presented with an award (a wind-up bunny) for the most kilometers walked on stage while giving a lecture at a conference in Germany. When recording a podcast, I find it difficult to keep a consistent distance from the microphone. I simply have difficulty speaking if I can't move freely. Maybe this is because I learned to speak in Italian, a language rich with gestures. So rich, in fact, that a traveler to Italy can purchase an Italian gesture dictionary.

Gestures differ across languages—even a hand signal that may seem very basic to someone from the US, like holding up an index finger to signify "one" might get you *two* beers at a bar in parts of Europe. There are many travel-tip resources available cautioning travelers about seemingly innocent gestures that might get you in trouble.

There is a baseline difference in the rate of gesturing among languages. Mandarin speakers, for example, gesture less, on average, than English speakers. However, Chinese-English bilinguals increase their gesture rates while speaking Mandarin, the gesture rate of one language influencing the other.[42] The choice of *which* word to supplement with a gesture can vary across language. As an English speaker, I would probably accompany the phrase "go outside" with a gesture indicating the preposition "outside." A Spanish speaker, however, is more likely to accompany that phrase with a gesture expressing the verb "go." A Spanish-English bilingual, when speaking English, would likely maintain the verb-centered "go" gesture.[43] As a rule, gestures seem to be more "sticky" than spoken language.[44]

What do we know about how *emotions* are expressed and felt in a bilingual's languages? The way we express emotion is weighted differently across languages. For example, if there is a mismatch between voice and facial expression, Japanese listeners weigh their evaluation of emotion more toward the voice. Dutch listeners assign more weight to the facial expression.[45] Emotions are, broadly speaking, *felt* differently in the two languages of a bilingual.[46] It is generally agreed that emotions are felt less strongly when speaking in one's second language. For this reason, a bilingual may deliberately switch to his second, less emotionally laden language when needing to make a rational decision.[47]

From a biological standpoint, bilingualism affects how we sense and how we think. It features in how we express and perceive emotion, and it affects the movements we make while speaking. Speaking

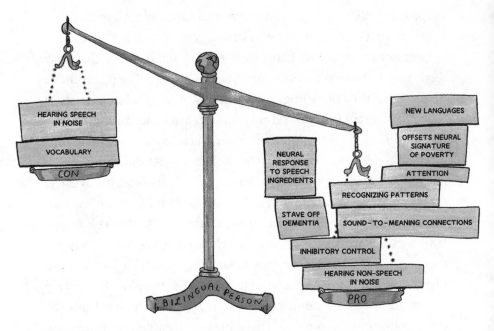

Figure 9.6
The benefits of speaking two languages outweigh the disadvantages.

two or more languages offers rich linguistic, cognitive, and gestural possibilities for the sounds we hear and produce. The bilingual sound mind differs from a monolingual sound mind, in line with my thesis that our lives in sound shape who we are.

Taken as a whole, the downsides of speaking two languages are outweighed by wide-ranging and sometimes profound advantages (figure 9.6). A superpower indeed.

10
Birdsong

Perhaps what they're doing with all their vocalizing is what humans do when we get together to play music; perhaps it simply builds cohesion, a sense of group identity. If sheer vocal expression is older than language, perhaps art preceded words. If art is older than words, perhaps that is why the world is so full of beautiful display. Maybe these birds are all artists, masters in the art of being parrots, playing their parrot jam sessions.

—Carl Safina

Birdsong has a lot to teach us about ourselves and the creatures with whom we share the planet.[1] You may wonder why I am giving the song of a bird more attention than any other environmental sound like a woodpecker tap, a cricket chirp, a cat meow, or, for that matter, a babbling brook or a traffic jam.

It is worthwhile to give birdsong its due for several reasons. Historically, and probably prehistorically, humans listened to birds for a practical reason. The sound of birds—whose calls and songs conveniently fall squarely in the human range of hearing—would tell our ancestors that a given area was fecund. We learned that an environment able to support a healthy bird population likely would be a good place to settle. Listening to birdsong helped shape our human geography.

Second, from a biological perspective the songbirds' vocal apparatus is similar to ours. The brain structures involved in making and processing songs are broadly analogous to those in humans, including the efferent pathways feeding back from the cortex to the thalamus and midbrain.

Third, songbirds share with us the capacity for vocal learning—the rare imitative ability at the heart of language and communication.

Fourth, birdsongs follow developmental time courses, sound ingredients, and even grammars that are suspiciously like human language.

Fifth, like much human singing, it's all, or at least mostly, about sex.

Finally, birdsong is beautiful.

So, for all these reasons, songbirds and their songs can help us understand the sound mind.

Which Birds Sing?

Most birds make some sort of sound. However, not all birds are *songbirds* and not all birds sing. Chickens and ducks, woodpeckers and hoopoes, owls and doves, quails and cranes make various *calls*. However, these are not songs and their makers are not songbirds. Songbirds include wrens, robins, cardinals, sparrows, larks, swallows, orioles, finches, and many others—some 4,000 species.

Songs are used primarily by male songbirds to attract mates or to advertise their territory (with the purpose of attracting a mate with his swanky patch of forest). Songs tend to be longer in duration than calls because to a potential mate, a long song is more attractive than a short song. This is thought to be due, in part, to the implication that a long, well-developed song could only arise if the male singer had a healthy development in the face of nutritional and other stresses faced during the early part of life when the song was developed.[2] A call, in contrast, is almost always shorter in duration and less complex than a song. It will often be more of a "peep" or "squawk" in quality and it typically serves a nonattraction function. It may be used as a warning or to coordinate location information among members of a flock. Or, in young birds, to say "feed me!" Another key difference is that *a song must be learned.*

Mechanics of Birdsong

The vocal apparatus in the songbird has a different name than its human counterpart, the larynx, but it shares many characteristics. The avian equivalent is the almost-rhyming syrinx. But, like flight popping up independently in various corners of the animal kingdom—the bat, the bird, the insect[3]—the larynx and the syrinx are evolutionarily independent.[4] The syrinx seems to have arisen de novo, not through evolutionary adjustments to existing traits or structures in songbirds' ancestors. It is positioned low in the trachea where bronchi from the two lungs meet, unlike the larynx, which is situated high in the trachea, well north of the bronchial merge. But very much like the larynx, the syrinx has folds that vibrate as air from the lungs rushes past, producing sound. The tension of the folds determines the vibrating frequency and thus the pitch of the note.

Birds are one-up on humans, though. By virtue of the syrinx's location at the fork of the two bronchi, there are *two* sets of vocal folds activated by air coming from each lung. Often, they act together, but birds can activate them separately, either each in turn or simultaneously. Higher notes are produced with the vocal cords on one side of the syrinx and lower notes in the other. Songbirds can switch sides seamlessly. The cardinal, for example, creates a high-to-low frequency sweep that starts on the right side and switches undetectably to the left midstream. Birds can even sing duets with themselves, with different notes simultaneously sounding from the two sides.[5]

This calls to mind the haunting melodies of Tuvan throat singing, where more than one pitch is sounded simultaneously. However, the mechanism is quite different. A Tuvan throat singer produces a *single* fundamental frequency and uses exquisite control over the mouth, tongue, and lips of the vocal tract to selectively emphasize some harmonics and suppress others nearly completely. A speech sound has a full complement of harmonics, and we use our articulators to emphasize a handful of them—that is, create bands of frequencies—to form the desired vowel. Tuvan throat singers turn that principle up to eleven, having developed profound control over the whole range of audible harmonics, emphasizing and deemphasizing a much wider range of harmonics to form the high notes while the base note, the fundamental frequency, stays the same.

Birdsongs can be very loud. The nightingale can reach 95 dB, well into the exposure range that would mandate hearing protection in the workplace.

Another characteristic of birdsong is the ability to move from note to note extremely quickly. This shows up as a trill. The fast warble of the trill comes from the superfast syringeal muscles changing position in 4–10 ms,[6] far faster than nearly any other muscle is capable of, and only matched in a few other circumstances in the animal kingdom, such as in the rattle of the rattlesnake. Speaking of fast, songbirds also take "mini-breaths" that enable their songs to sustain for minutes without a pause.[7] Such durations would be impossible based on the lung capacity and normal respiration rate of a songbird, which is one to two-and-a-half breaths per second. These mini-breaths are on the order of four-hundredths of a second[8] and are synchronized with each note, enabling a nightingale to sing, in one recorded case, for twenty-three hours without a break, far exceeding the length of even the most ambitious opera aria like Brünnhilde's twenty-minute "Immolation Scene" in *Götterdämmerung*.

Birdsong and Speech

Aside from the mechanics of production, there are acoustical similarities between birdsong and speech. Both speech and birdsong consist of ordered strings of sounds separated by brief silent intervals.
A single birdsong note is roughly analogous to the phoneme, the
smallest unit of speech. With this raw material, notes or phonemes
are strung together to make motifs—as many as 100 in the case of
the woodlark,[9] 180 in the nightingale[10]—or words; motifs or words
are strung together to make songs or sentences with the sequencing of the motifs within the song ever changing. Thus, like speech,
birdsong contains information in multiple timescales, from the tens-
of-milliseconds note to the hundreds-of-milliseconds motif to the
seconds- or even minutes-long song that follows rules of syntax.

Like speech, birdsong has dialects. Songbirds have dialects or
accents even when speaking the same "language."[11] A given species
in one region will have a song that sounds a bit different from those
in another region, even though they are recognizable as belonging to
the same species. These differences in dialect are of vital importance
to the female bird. She is much less interested in the "accented"
songs of a visitor from another region than she is in the songs of a
male who shares her local dialect.[12]

The fundamental frequencies, the harmonic enhancements and
suppressions, and the often incredibly fast changes in birdsong are
best illustrated by the spectrogram. Not just a tool for the sound
specialist, the spectrogram for years has appeared alongside pictures
in bird identification books for hobbyists. Similar to reading sheet
music, reading a spectrogram lets us visualize note durations, frequencies, and movement at a glance, so that a quick "match or not-
a-match" judgment can be made by a birder who is trying to decide
whether a new "little brown job" can be added to their life list. A
simple example spectrogram of a human melody can be seen in
figure 10.1 below its corresponding music transcription.

A birdsong may consist of one or more clear whistle-like tones, buzzy tones, or trills, which are notes repeated too fast to count (i.e., more than ten per second). These elements are combinable in rising or falling FM sweeps, or largely pitch-level sequences, and more than one sequence may comprise a single song. Some songs are fast and manic sounding; others unfold at a leisurely pace.

The song in figure 10.2 makes up part of a house wren's repertoire. The repertoire of a bird of a given species might consist of just one song—for example, the zebra finch or the white-crowned sparrow—or up to a thousand or more in the case of the brown thrasher. Here is where some of the analogies to human speech begin to break down. However large the repertoire might be for a given species, there is a distinct lack of flexibility. Human language has a boundless capacity to adjust, reorder, and evolve to convey meaning. While birdsongs may differ depending on context—advertising for a mate, declaring ownership of territory, pair-bond maintenance, etc.—they remain largely rote and inflexible. The semantic richness and infinite flexibility of human speech is lacking.

For these reasons, despite the many acoustical and anatomical similarities with human speech, birdsong *is* communication, but because it lacks the open-endedness and endless flexibility of human speech, it is generally not considered language.

Figure 10.1
Analogous to musical notation (top), a spectrogram (bottom) represents pitch on the y-axis and time on the x-axis. Though not depicted in this "Twinkling" example, dynamics (loud and soft) are rendered in the spectrogram by darkness of the lines.

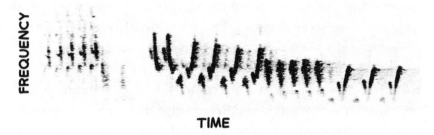

Figure 10.2
A complete 2.5-second song of a house wren. Spectrogram created from an audio sample downloaded from https://www.floridamuseum.ufl.edu/birds/florida-bird-sounds/.

Birdsong and Music

If birdsong is not language, is it music? After all, it's right there in the name: bird*song*. What characteristics define (human) music, and which are shared in birdsong? There is a lot of leeway in the identification of the defining elements of music. Pick any number between about five and twelve, and you will be able to find a corresponding website listing the "six" or "nine" or "twelve" elements that comprise music. Though numbers and terminology differ, they all basically boil down to familiar sound ingredients: melody, rhythm, and harmony (pitch, timing, and timbre). In addition, there is dynamics (intensity) and usually some word or phrase that expresses the idea of how it is all put together (e.g., structure, texture, form) that I will call composition here. Can we ascribe these elements to birdsong?

Pitch
The pitches over the course of a birdsong or within a motif are often precise and repeatable. They seem to occur at intervals we would consider consonant, like perfect fourths or octave jumps. The notes of a hermit thrush's birdsong belong to a harmonic series of an implied (i.e., not sung) base note.[13] There have been many other claims of musicality in various species' song pitches. The white-throated

sparrow and the ruby-crowned kinglet are said to produce consonant intervals.[14] The songs of the hermit thrush and the canyon
wren are said to conform to pentatonic and chromatic scales,[15]*
respectively, although there has been little actual acoustical analysis
to back up these claims.[16] On the contrary, a large sample of north-
ern nightingale wren song samples revealed that birds are no closer
than chance at producing notes that fall into any human construct
of the scale, whether diatonic, pentatonic, or chromatic.[17] Nevertheless, it is interesting and thought-provoking to view musical
transcriptions of birdsong, like the one in figure 10.3. The melodies of birdsongs have inspired, evoked, or been explicitly featured
in compositions by Vivaldi, Hayden, Vaughan Williams, Bartók,
Beethoven, Mozart, Frescobaldi, Schubert, and Messiaen.[18] Respighi
incorporated an actual recording of a nightingale into the third
movement of his "Pini di Roma."

Timbre, Timing, and Intensity

Although probably not strictly under the bird's control, a wide range
of timbral qualities can be heard in their songs. Baptista and Keister
have noted timbral similarities between a number of birds and musical instruments: the song of the Australian diamond firetail finch
resembles an oboe, the common potoo a bassoon, the strawberry
finch a flute.[19] The same authors also tick off examples of *accelerando*
and *ritardando* (timing) and *crescendo* and *diminuendo* (intensity or
dynamics) in the songs of a number of species. Birdsong shares some
similar rhythmic patterns with human music as well.

Composition

Listeners have noted final flourishes, cadences, bridges, and descending cascades akin to piano *glissandos* (FM sweeps).[20] Birds with large

---

*The diatonic scale is the familiar seven-step mix of half and whole steps that goes
"do-re-mi-fa-sol-la-ti[-do]." The pentatonic scale has five steps "do-re-mi-sol-la[-do]."
The twelve-step chromatic scale fills in whole-step gaps of the diatonic scale.

Figure 10.3
A transcription of a veery thrush's song by Tony Phillips. From http://www.math
.stonybrook.edu/~tony/birds/.

repertoires string them together in movements, sometimes evocative
of a theme and variations. Pairs (or more) of birds sometimes sing a
call-and-response canon. Practitioners of this behavior include the
Socorro mockingbird and the marsh wren.

Brainvolts alumnus Adam Tierney, who to my recollection never
weighed in on whether he considered birdsong music, nonetheless used birdsong to test a hypothesis about human songs. Human
songs have a noted preponderance of three characteristics: (1) close
pitch spacing between notes, (2) melodic contours that are either
descending or Λ-shaped (rather than ascending or V-shaped), and
(3) a tendency of having prolonged notes at the end of a phrase.
Adam hypothesized these characteristics are due to motor constraints and not innate or cultural preferences. His approach was to
tally the prevalence of these features in birdsong because birds have
similar motor constraints to humans due to the analogous anatomy
involved in song. After an analysis of a large corpus of birdsong
recordings, he found that birdsong shared the same three characteristics, suggesting a physiological basis for the forms that human
songs take.[21]

Nevertheless, the question of whether birdsong is "music" remains
very much up to the individual. Composer and zoömusicology
researcher Emily Doolittle writes[22] that she once made a list of ways
in which birdsong did *not* resemble human music. Her list included

things like "No overarching structure," "No harmonic relationship between different motifs," and "Arbitrary alternation of sound and silence." She then showed her list to fellow composer Louis Andriessen, who said, "That sounds like Stravinsky!"

Vocal Learning

Vocal learning is distinct from *auditory* learning. Your dog is an auditory learner—she understands "sit" and "walk" just fine, but she will never learn to say them. There is more to the distinction, though. Dogs and many other animals use auditory learning to learn to vocalize *appropriately*. An animal must learn, for example, that a given innate call should be used only to signify alarm; misuse would inappropriately alarm members of the social group. But these animals did not learn their alarm call (barking, growling, whining . . .) by explicitly modeling it on others' calls. They already instinctively knew how to bark, growl, or whine. In contrast, songbirds, though armed with instinctual abilities to *vocalize*, cannot learn to *sing* without modeling this behavior from another bird's song and molding their vocalizations into songs through practice. This process is vocal learning in action.

Vocal learning relies on listening, memory, and imitation and requires good motor control of the muscles powering the vocal apparatus that many nonvocal learners lack. Much of our—humans'—learning comes from imitation; and where speaking is involved, "much" becomes, for all intents and purposes, "all." As with humans and the other handful of vocal learners, the process of song learning in songbirds has the following four characteristics: *imitation, auditory-motor feedback, sensitive periods,* and *brain lateralization.*

Imitation

Any songbird species has its signature song (or, in some cases, songs): the "cheerily, cheer-up, cheerily" of the robin, the "potato chip" of the goldfinch, or the "bob-white" of the, er, bobwhite.[23] The young male bird—studies on birdsong traditionally focused on males, but this is beginning to change—makes his first vocalizations, recognizes how they differ from his tutor's model songs, and makes the necessary adjustments until it matches. During the process, the tutor may adjust his song by introducing additional repetition or increasing pauses between motifs, in much the same way a parent will adjust their manner of speaking to an infant.[24] If you isolate a young bird from its father and other males who also may serve as song tutors, he does not learn the song characteristic to his species.[25] Chaffinches raised in isolation from a young age develop abnormal songs, albeit with some species-appropriate motifs.[26]

Songbirds have a predisposition for learning their own species' song, but this is overshadowed by their need for a live tutor. Naïve young birds' heart rates increase[27] and their auditory systems perk up[28] at the sound of their species' song compared to one of another species, and they will selectively vocalize in response.[29] In fact, this predisposition can be seen in songbirds who can be readily trained to perform a task when the reward is hearing a recording of a song of their own species. Rewards of another species' song are unmotivating.[30] However, a young bird does not do a very good job of learning his species' song from a tape recording. He will be more successful imitating a live bird of another species.[31] So, as with human babies, who at a young age develop preferences for the sounds of the language spoken by their parents,[32] and who cannot learn language as well from television or audio recordings,[33] there is a strong social-interaction component.

Recognizing melodies transposed in pitch (fundamental frequency) is something humans can do with ease. It enables us to sing together. It is why wolves adjust the pitches of their howls when other wolves join in.[34] Birds, however, do not recognize a transposed melody.[35] Instead, birds rely on harmonics, that is, the shape of the spectrum, to recognize melodies, much as we recognize which word was spoken from bands of harmonic energy (remember figure 1.6?), not the pitch of the voice speaking them.[36] Unlike humans, wolves, and mice,[37] birds generally do not sing *with* other birds; they sing *to* other birds.

The imitative style of learning birdsong features the auditory system front and center. So what do we know about the role the avian auditory system plays in their songcraft?

Auditory-Motor Feedback

In songbirds, hearing a tutor's song and learning to produce it involves the transparently named "song system." The song system involves pathways that entail auditory and motor brain regions and, ultimately, the muscles that control the syrinx. Crucially, some of these cortical and subcortical connections, within the auditory system and between the auditory system and the syrinx, are absent in nonvocal learners, including even species of birds that are not songbirds. Damage to portions of a bird's song system, including strictly "auditory perception" parts, damages the bird's ability to learn and produce songs in much the same way as damage to certain auditory areas of the human cortex (e.g., in stroke) creates aphasia or other conditions that hamper a human's ability to speak fluently.[38]

As in humans, neurons in the songbird auditory cortex initially respond to any sound, but with experience will tune to the sound of a tutor's song.[39] A "comparator" circuit at the intersection of the auditory and motor systems continues to refine the learning bird's song until the difference between his output and the tutor's input effectively vanishes,[40] and a song has been learned. If a bird is deafened before his vocalization practice phase, this comparison cannot be made and the songs that emerge are highly abnormal.[41]

While most birds stick to their species' song(s), mockingbirds are a well-known example of a bird who excels at imitating other species' songs. Another champion imitator is the lyrebird.* The

---

*Please look up a wonderful BBC Studios video clip with David Attenborough narrating as a lyrebird imitates a car alarm, a chainsaw, and a camera shutter (complete with an automatic film advance mechanism!).

tight coupling between movement and sound was evident when the coronavirus lockdown resulted in vastly reduced human-made noise. During this period, birds produced more technically and motorically challenging songs.[42]

Sensitive Periods

Songbirds develop their songs at a critical period in their development. First, the bird listens to and memorizes its tutors' song.[43] Then the bird matches its own singing to that of its models. Like human babies, songbirds go through a period of babbling, known as subsong. Using auditory feedback, they gradually change subsong into plastic song. The plastic song phase involves not just practice but also a winnowing process. A bird might try out a number of songs from a variety of tutors until eventually, after a sometimes-long process that might require tens of thousands of practice songs, he arrives at his adult-like crystallized song (figure 10.4).[44] And then that's pretty much it. Once the song has crystallized, it stays the same throughout a bird's adult life, regardless of the amount or variety of subsequent song exposure.* The listening and memorization stage usually occurs in the first few months of life. The vocalization, refinement, and crystallization stages occur at the onset of sexual maturity. Isolating birds from tutors, deafening them, or otherwise interrupting the natural process during these critical learning phases results in songs that differ from their tutors' songs. Songs may be very limited or consist of a sequence of disorganized sound elements.[45] The fact that birds raised in isolation—that is, away from any tutors—produce songs at all suggests that some elements of birdsong are instinctual and others are learned.

Brain Lateralization

Why certain brain functions are lateralized can be likened to the need for people with different outlooks for the same job.[46] If your

---

*Exceptions to this rule exist, of course. The canary, for example, repeats this learning process over and over again as an adult, arriving at a new song each spring.

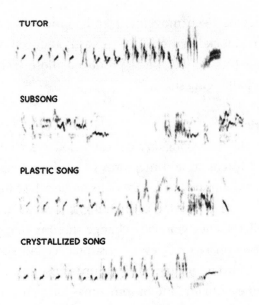

Figure 10.4
A young chaffinch eventually produces a song that matches that of his tutor (top). Adapted from M. Naguib and K. Riebel, "Singing in Space and Time: The Biology of Birdsong," in *Biocommunication of Animals*, ed. G. Witzany, 233–247 (Dordrecht: Springer Science+Business, 2014).

left visual brain is focused on foraging for food, then your right visual brain can be monitoring for predators. In humans, the right and left cerebral hemispheres play distinct roles in language. There is a brain-hemisphere difference in songbirds while listening to birdsong. In the zebra finch, for example, the right forebrain is more responsive to their species' songs.[47] In songbirds, too, the two hemispheres process sound distinctively.

Sex and Song

In most cases, the male bird is the singer and the female bird chooses her mate based on the song she prefers. Does needing to be a discerning listener tune the female's auditory brain differently from the male's? The song of a male is affected by visual signals from the female. A male will vary his song until a female decides she likes a particular variant. She will signal this with a visual display such as a wing stroke, which is a rapid movement of the wing away from the body. A male recognizes this signal as a sign of approval and will then repeat the song variant that elicited it, often leading to copulation.[48] I'd like to know more about the sound ingredients a female bird finds sexy. How do the ingredients they prefer shape the kinds of offspring that materialize and ultimately the development of the species overall? How might sound processing differences in male and female birds compare to the sex differences in sound processing observed in humans?[49]

The brain regions that are active during singing are context dependent. When a male zebra finch sings with no audience, regions involved in song learning and self-monitoring are active. However, when a female is listening, those particular brain regions are quiet. Birds, like humans, seem to distinguish between practice and performance. Like humans who use different brain regions when improvising or playing according to a score,[50] birds utilize context.

Daily and seasonal variations in song production are largely hormonally controlled. Castrating a male keeps him from singing, and testosterone induces singing in females.[51]

Females of some species sing duets with males.[52] In other species, like the stripe-headed sparrow, females outsing the males.[53] A survey of over a thousand songbirds found that females sing in 64 percent of them,[54] though in no species did *only* the female sing. The species with female singers tend to be those with brightly colored plumage. Females seek the most melodious and beautiful mates. The correlation between bright plumage and singing suggests these traits may have evolved together. According to Carl Safina, "Beauty—for the sake of beauty alone—is a powerful, fundamental, evolutionary force."[55]

11

Noise: Stop That Racket, It's Hurting My Brain

Can we have a moment of silence, please?

My Italian home base, Trieste, is near the Dolomite mountains. I have hiked there all my life. One recent spring, after a long ascent, my cousin Lucio and I sat at the top of the world, looking at the peaks and valleys around us and listening. I lay back in the grass. After about ten minutes of just being, I said something to Lucio. When I broke the silence, the loudness of my voice was jarring. The lack of noise called for a recalibration of listening.

The sound mind accomplishes the herculean task of turning air movement into sensation, sound into meaning, on a routine basis. But what about the sounds that serve as obstacles to our extracting meaning from intended sounds? One of the impediments that gets in the way of our well-tuned auditory system—and it is a big one—is noise. I'm talking about noise in its usual sense—unwanted sound outside the head. But I also want to talk about noise inside the head—the conditions that impede the sound mind from efficiently doing its job and what, if anything, we can do to combat it.

What Is "Noise"?

Noise has an interesting etymology in English. It comes to us from
the old French word for quarrel or dispute. It also shares a root with
the Latin "nausea," which means seasickness—a gut reaction, literally, to something negative. Noise is an unwanted sound, a negative
and perhaps damaging sound.

Since antiquity, sound has been recognized as a destructive force—
the walls of Jericho were said to be brought down by loud sounds.
The Sirens' song, though beautiful, lured seafarers to their demise.
Contemporary uses of sound as a force include directed ultrasonic
sounds for crowd control and loud high-pitched noises piped into
public or private spaces to deter animals or teenage loiterers. A typical adult cannot hear these high-frequency sounds, yet might they
nonetheless damage hearing? We have evolved to react to unexpected sounds. Our ancestors managed not to be eaten because
sound alerted them to the presence of predators. Unexpected sounds
continue to alert us, though rarely signal a matter of life and death. A
phone rings, a door unlatches, a toilet flushes, a dog barks, an alarm
goes off, a shout wafts through the window. These may be noises in
the sense that they are not particularly desired. But these are not the
noises I am talking about here.

I am also not talking about loud noises that are blatantly damaging to the ~30,000 specialized hair cells in the ear. It is well
documented that these hair cells can be damaged by exposure to
sounds at high decibel levels. The National Institute for Occupational Safety and Health (NIOSH) has published guidelines for the
maximal amount of noise one should be exposed to (figure 11.1).
For example, if the ambient noise level is 100 dB,* a safe exposure

---

*dB means decibel. Typical activities where 100 dB is encountered include using a
hand drill, riding a motorcycle, riding the subway, using a leaf blower. A garbage
truck with its compactor operating is about 100 dB, as is a jet flying overhead at

NIOSH'S PERMISSIBLE NOISE EXPOSURES	
SOUND INTENSITY (DB)	**DURATION**
82	16 HR
85	8 HR
88	4 HR
91	2 HR
94	1 HR
97	30 MIN
100	15 MIN
103	7.5 MIN
106	3.75 MIN
109	<2 MIN
112	~1 MIN
115	30 SEC

Figure 11.1
Guidance for noise exposure time by intensity.

dose is a mere fifteen minutes, as exposures beyond that duration represent an increase in the probability of eventual hearing loss. Despite these guidelines, noise-induced hearing loss remains the most common occupational hazard in the United States.[1]

The type of sound that NIOSH is concerned about, the loud sounds, can cause hearing loss *in the ear*. My focus here is on moderate-level noise—the type usually thought of as "safe" noise because it is not known to damage the ear per se. In short, I'm talking about the difference between harming our ear and *harming our brain*.

1,000 feet. Playing some (even nonamplified) musical instruments, attending concerts, or listening to loud music with personal music players also can expose you to this intensity level. Do you cumulatively do any of these activities for more than fifteen minutes in a day?

The Biological Impact of "Dangerous" Noise (Damage to the Ears)

What do I mean by the "ear" type of hearing loss? The term "hearing loss," along with hearing impairment and deafness, is usually boiled down to numeric *thresholds*. Thresholds are evaluated with the familiar "raise your hand when you hear the beep" test. A hearing specialist tests your ability to detect tones across a range of pitches considered essential for hearing speech and assigns a threshold, or the softest intensity level at which you can detect a sound. By convention, anything below 20 dB is considered "normal." Increasingly higher (worse) thresholds are termed moderate, severe, and profound hearing losses. Hearing loss can be "flat" (the same threshold across frequencies) or sloping (worse thresholds at high than low pitches), or can come in some less-common configurations. Hearing loss measured this way is an evaluation of whether the *ear* is doing its job.

Increased hearing thresholds caused by noise exposure can affect our personal and professional lives. My son recently took his car to a shop because he was hearing a subtle but worrisome noise he guessed was coming from the transmission. The mechanic took the car on a test drive, heard nothing concerning, and sent my son away with the assurance he had nothing to worry about. A week later, the car had a transmission failure. Years of working in a noisy garage had likely damaged the mechanic's hearing.

Hearing protection is important for a wide range of industrial, factory, and construction workers. It also can be vital for musicians, but is often overlooked. A symphony orchestra often approaches 100 dB, and brass and percussion instruments can far exceed this level during a loud passage. A violin is not extremely loud, but its F holes are inches away from the left ear. Violinists routinely are found to have worse hearing thresholds in their left ear than their right. Most hearing protection dampens the higher frequencies disproportionally. Newer designs have emerged that do a good job of reducing sound levels equally across the frequency spectrum. This and other topics relevant to hearing preservation in musicians are well covered in a book called *Hear the Music: Hearing Loss Prevention for Musicians*.[2]

So we know that exposure to loud noise can increase your hearing thresholds. This is an important issue, but not the focus here. An interested reader will have no difficulty pursuing the topic of ear-damaging noise on websites hosted by the Centers for Disease Control and Prevention and the National Institute of Deafness and other Communication Disorders.

The Biological Impact of "Safe" Noise (Damage to the Brain)

We need to be less cavalier about the day-to-day commotion that surrounds us in our raucous world. These noises do not meet or exceed the generally accepted threshold of "unsafe." They are not novel and alerting; rather, they are ongoing and have generally consistent acoustic properties over time; hence, they do not convey much information. These are the sorts of sounds most would consider "background noise." For this reason, we tend to ignore them. We tune them out. But are we really tuning them out, or are we simply living our lives in a constant state of alarm? We have all experienced not noticing a sound until it goes away. Often it is an air conditioner or an idling truck. The air conditioner cycles off or the ignition is cut, and suddenly we "hear" the silence. And we sigh in relief. We momentarily revel in the peace until it starts up again or is replaced by the next aural annoyance. If our *ears* are not being damaged and we can mostly tune it out, should these noises concern us? Science tells us we should indeed notice it and be concerned for the sake of our *brains*.

Difficulty understanding speech in noise after exposure to moderate levels of noise can emerge in people with normal hearing thresholds. Moreover, a noisy environment has many underrecognized negative impacts that have little to do with hearing per se. Chronic noise exposure—for example, such as might be experienced by individuals who live near an airport—can lead to an overall decrease in perceived quality of life, increased stress levels along with an increase in the stress hormone cortisol, problems with memory and learning, difficulty performing challenging tasks, and even stiffening of blood vessels and other cardiovascular diseases.[3] According to the World Health Organization noise exposure and its secondary outcomes such as hypertension and reduced cognitive performance are estimated to account for an astounding number of years lost due to ill health, disability, or early death.[4]

Noise disturbs learning and concentration. Students attending public schools in New York City had markedly different reading outcomes depending on whether their classroom was on the side of the school that fronted a busy elevated train track or on the other side of the school, which was shielded from the train noise.[5] Students on the noisy side lagged three to eleven months behind their peers in reading. In the wake of these findings, the New York Transit Authority installed rubber padding on the railroad tracks near the school and the Board of Education installed noise-abatement material in the noisiest classrooms, together reducing noise levels by about 6–8 dB. The reading-level difference soon vanished.[6]

The effect of noise is not limited to auditory or language tasks like reading. In one experiment, subjects were asked to track a visual target, a moving ball, on a computer screen with a mouse. Meanwhile, other balls were simultaneously roving around on the screen. Participants who had experienced long-term noise exposure as part of their occupation had a more difficult time with the task, especially when the task itself was accompanied by random noises; they were slower and unable to keep as close to the target ball.

In *Why We Sleep*,[7] UC Berkeley sleep scientist Matthew Walker calls the lack of proper sleep "the greatest public health challenge we face in the twenty-first century." Sleep is becoming more recognized as crucial for our health, as it affects our cardiovascular system, our immune system, and our ability to think. Noise is one of the biggest culprits keeping us from a good night's rest. Noise—even at fairly low sound levels—has a harmful impact on quantity and quality of sleep. Noise keeps us awake longer and awakens us earlier. While sleeping, noise in the environment affects the quality of sleep, prompting body movements, awakenings, and increased heart rate. Traffic noise can shorten periods of REM (dream) and slow-wave (deep) sleep, and diminish one's perception of the restfulness of a night's sleep.[8]

In our waking lives, the insult of "safe" noise to the sound mind can be especially pernicious for children. Children are masters of language learning. Parents are gobsmacked at the short amount of time that elapses between observing their child say their first word to their speaking in full sentences. Sound to meaning connections are formed with great rapidity. Children cannot help learning the languages they are exposed to—even more than one. But what if the sounds children are exposed to at this critical age are meaningless?

This question is difficult to address in humans because it is impossible to control noise levels adequately in a real-world setting. However, we can answer questions like this in animal experiments. By controlling the duration, intensity, and quality of sound exposure, it is possible to get a direct look at how the electrical signals—the currency of the nervous system—in the brain are affected. Just what happens to our sound minds when we are exposed to "safe" noise? And are these effects transient or permanent?

Typically, by adulthood, the auditory cortex of a rodent is organized tonotopically. However, early in life, low- and high-pitch sounds have not yet settled into their cortical homes. Developing rodents were raised in an environment with continuous 70 dB noise. For reference, the NIOSH table does even not go that low; 70 dB is considered a "safe" level of noise. By the time they reached maturity, their auditory cortices were still undifferentiated in terms of tonotopy; the low- to high-pitch gradient had not formed (figure 11.2).[9]

This raises concerns for human babies who spend time in an environment we might consider noisy but not "damagingly" noisy, like a neonatal intensive care unit (NICU).[10] What might happen to a premature baby's auditory cortical organization as she listens to the beeps and clatters of medical monitoring systems, ventilators, and pagers rather than typical intrauterine sounds like rhythmic heartbeats, digestive noises, and the filtered voice of her mother that she would still be enjoying if born full term? Preterm infants can have a host of developmental challenges, including language and cognition, that may be exacerbated by this early noise exposure.[11]

Scientists have introduced measures to mitigate the noisy NICU atmosphere.[12] In one study, the sounds of the mother's heartbeat and voice were piped into the incubator. Babies with exposure to these "good" sounds along with the bad had a more fully developed auditory cortex than the infants who heard only the bad sounds.[13] Live music performed in the NICU also stabilized babies' heartbeats, reduced stress, and fostered sleep.[14]

Cortical map disorganization need not be permanent. In rodents whose tonotopic maps were disorganized by noise, once the

Figure 11.2
"Safe" noise scrambles sensory maps.

noise was removed, the tonotopic organization of the cortex resumed afresh.[15] Similarly, after noise damage, cortical map disorganization can be minimized by exposure to an enriched auditory environment[16]—reminiscent of the positive effect of enriched sounds for babies in the NICU. The sound mind is constantly reinventing itself.

Does susceptibility of the auditory brain to "safe" noise diminish later in life? Adult animals were exposed to "safe" levels of noise, again in the 60–70 dB range, for several weeks. Their hearing thresholds did not change, but the way the auditory cortex responded to sound changed, reflecting a disorganized tonotopic pitch-processing mechanism.[17] The frequencies present in the noise took over the real estate in the brain that rightfully belonged to other frequencies. Thus, damage done by "safe" noise is not limited to sensitive periods during development, but can affect adults as well.

Given what we know about the biological damage of "safe" noise, we should reconsider the widespread use of noise generators, especially for the developing brain. Often used to keep people, including babies, from being awakened by household sounds, these devices, running eight or more hours at a time, might be blunting the sound mind and may have a long-term impact on our ability to effectively derive meaning from sound.

Noise Inside the Head

We should be concerned with noise *inside* the head as well as outside. Sounds do not arrive on a blank slate. Like the static between stations on the radio dial as we tune to the ball game or music channel, the brain is never quiet. There is always a base level of background activity—idle-state neural firing—that the sound mind needs to "tune" through. The neural response to sound must overcome this background electrical activity for the sound to be registered, so it is important that the idle-state activity not be overwhelming. We found an unexpected link between the size of this background brain activity and language development. Maternal education often patterns with the extent of language stimulation a child is likely to experience. It is also widely used as a proxy for socioeconomic status.[18] Divided on the basis of maternal education, children of more educated mothers have a lower level of background activity—a less noisy brain. These children also had more precise processing of sound ingredients (figure 11.3).[19] That is, learning to make effective connections between sound and its meaning is likely to lead to a clearer signal and reduce the background neural activity so that effective and precise sound processing can take place.

Families of low socioeconomic status risk being exposed to a less rich linguistic environment[20] *and* tend to live in noisier neighborhoods. Maybe the background noise level of the brain becomes amped up from long-term exposure to traffic and train noise, proximity to industrial sites, and crowded housing that track with lower

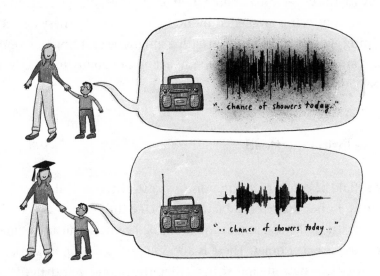

Figure 11.3
The spontaneous background noise of neural firing is greater in individuals with lower levels of maternal education—a proxy for income level.

income.[21] Support for this interpretation comes from animal experiments in which noise exposure led to an increase in spontaneous brain noise—a sort of hyperactivity of the brain—in the auditory midbrain and cortex.[22] So noise inside the brain can be caused by noise outside the brain. A higher baseline level of internal noise is competing for "headspace" with important sounds like speech. A lifetime of noise exposure and linguistic understimulation, in a vicious cycle, can compromise the ability to make sense of sound.

Tinnitus (the pronunciation of which even specialists cannot agree on—is it TINN-i-tus or ti-NITE-us?) is another example of "noise inside the head." Most commonly expressed as "ringing in the ears," it can also be a hiss, a buzz, or a hum. But the sound is not coming from an external source—it is generated internally. Tinnitus can be temporary, such as after attending a loud concert, or it can be chronic, leading to stress, depression, fatigue, and trouble concentrating. Chronic tinnitus can occur for many reasons that are poorly understood and frankly unknown.[23] Often, tinnitus is accompanied by hearing loss, and hearing loss brought about by noise, in particular, is a chief culprit. So in tinnitus we see a direct link between noise *outside* the head and noise *inside* the head.

Even when there is hearing loss, the source of tinnitus is the brain. Tinnitus, in ringing form, usually matches the frequency of the sufferer's hearing loss. If you have sustained a hearing loss (increased thresholds) at 2,000 Hz, the ringing will occur around 2,000 Hz. This is an auditory analog of phantom limb syndrome in which an amputee feels pain in their missing limb. It may be auditory neurons firing randomly despite not getting input from the ear. The auditory brain is always searching for stimulation; when sound is absent, the brain can make it up. Perhaps this is also why there is increased neural noise in children with linguistic deprivation.

White-noise-generating devices are sometimes used to distract tinnitus sufferers from the unwanted ringing. However, white noise might actually worsen tinnitus by exacerbating the abnormal function of brain centers that contributed to the problem in the first place.[24] If sound is used therapeutically to mask tinnitus, more meaningful sounds such as music, waves, or wind are likely to be more beneficial than invariant noise generators.

Hyperacusis and misophonia are an oversensitivity to sounds of moderate sound levels. These conditions often co-occur with tinnitus but can be experienced on its own. Tinnitus, hyperacusis, and misophonia provide striking examples of the auditory system's communication with our emotions. Attention to unwanted sounds, along with the negative emotions and stress it causes, drives a feedback loop that can worsen these conditions.[25] There is some hope that by therapeutically stimulating the emotion-gating limbic system the brain can be taught to reduce these disruptions to the sound mind.[26] Tinnitus, hyperacusis, and misophonia are thought to arise from a hyperactive auditory midbrain and cortex likely caused by a malfunctioning efferent feedback system that is failing to perform its inhibitory job.[27]

Biological Impact of Noise in the Environment

One of the properties of sound is its ability to operate over a distance. Seafaring Inuit and Tlingit have traditionally used their hearing to detect the sounds of whales beneath the hulls of their boats. Tutsi and Hutu people can hear the low-frequency communication of elephants.[28] But these abilities are beyond the grasp of most who have not learned to listen to sound details so acutely.

One of the reasons we are unable to listen carefully—and have turned into such a visually biased society—is because of all the noise. In *One Square Inch of Silence* Gordon Hempton tells us how his attention gradually shifted from listening to looking as he hiked the 150 miles into Washington, DC.[29] As he approached the capital, air traffic noise became almost continuous. By Hempton's reckoning, there are only *twelve* places in the world where silence can be experienced for fifteen consecutive minutes. And, to be clear, "silence" does not mean lack of sound. Rustling leaves, trickling streams, and singing birds count as silence in Hempton's judgment. Rather, he is speaking of automobile traffic, airplanes, farming machinery, leaf blowers, and other human-made sounds. The cumulative extent of human-made noise is so extreme that seismographs designed to detect tectonic disturbances and earthquakes can detect it.[30]

What is the impact of noise on animals? Birds, frogs, and even whales increase the loudness of their calls, change their call rates, or change the sound quality of their calls as their environments get more polluted by noise.[31] Song sparrows in urban areas shift the pitch of their calls from about 1,000 to 2,000 Hz to avoid the ambient city noises, which peak below 2,000 Hz.[32] Many of us noticed an increase in the loudness of birdcalls and songs when the 2020 coronavirus pandemic forced a severe curtailment of human-created noise. However, during this time, birds actually *reduced* the loudness of their songs in response to the diminished din of human activity while doubling the distance they could be heard. At the same time, they increased song intricacy.[33] Whales, if the noise pollution is severe enough, will simply go silent.* Moreover, the echolocation they rely on for navigation can be thrown off by ship's sonar, thought to be a cause of some beachings.[34]

The United States was privileged, over a hundred years ago, to have a president who was forward-looking about conserving our natural environment. Theodore Roosevelt established five national parks, eighteen national monuments, and over two hundred national

---

*Water, like air, is a vector for the transmission of sound; sounds conveyed by the movement of water are also subject to human-made noises. An interesting topic to pursue, beyond the scope of this book, is sound propagation in different media. The helium-balloon effect is an obvious one. Helium is less dense, and so sound travels at a faster rate and, correspondingly, our voice propagating through helium sounds higher. A 2017 episode of the podcast *Twenty Thousand Hertz* (https://www.20k.org) speculates about sounds on the planets of our solar system with their varying atmospheres.

forests, wildlife refuges, and game preserves. In his documentary, Ken Burns calls national parks "America's best idea." Roosevelt recognized the importance of preserving natural resources and the spaces that house them for future generations: "Of all the questions which can come before this nation, short of the actual preservation of its existence in a great war, there is none which compares in importance with the great central task of leaving this land even a better land for our descendants than it is for us."[35]

Too often, sounds are less appreciated than sights. We rally around causes that reduce visual pollutants and loss of forests, while there is a regrettable lack of awareness of the disruptive impact of noise on animal communication, mating, and indeed survival. We should regret the loss of silence and the impact it has on ourselves and other species.

What Can We Do about Noise?

Bianca Bosker, in the *Atlantic*, wrote about a man in Arizona who began to notice an omnipresent monotone hum at his house.[36] First attributing it to someone's pool pump or a carpet cleaner, he soon realized it was inescapable. Closing the windows or wearing earplugs wouldn't block the sound. After some sleuthing, he tracked it down to a data center a half mile away. The activity of our twenty-first-century electronic lives—the Instagram posts, the ATM transactions, the online purchases, the research involved in writing this book—all involve accessing data that have to be stored somewhere. So data centers, with acres of servers and their requisite heavy-duty cooling systems, churn out noise that Bosker calls the "exhaust of our activity." What can be done about noise pollution, and what can we do to mitigate its effects on our sound mind?

Job number one is to recognize noise as a powerful and detrimental force, even when it is not the type that makes us instinctively clap our hands over our ears. Noise fundamentally alters the sound mind and impacts our health. This biological evidence is underrecognized and underpublicized. Noise is virtually inescapable, so solutions are not easy. But reducing noise is something we can strive for. There are steps we can take to do this through our own behavior, through technology, and through enrichment. The first step is to simply become more aware of sound. Were you aware of the potential damage of exposure to "safe" levels of noise?

Download a sound-level measuring app on your smartphone and get a sense of your soundscape at home, at work, during your commute, and at the gym. Did you ever notice how noisy the gym can be? Between the overhead music, the clanking of barbells, the calls of the group instructor, and the brutal reverberation, it is a hostile aural environment. Ironically, we go to the gym for our musculoskeletal and cardiac health, but we may very well be damaging other aspects of our health. Maybe we don't have to slam that locker door so hard.

As we become more aware of the sounds around us, we can ask, "Is this necessary?" We can try to resist what the world is becoming and think before we passively accept the latest modern convenience. Does our clothes dryer really have to speak to us? Is it essential for the car to chirp or honk every time we lock or unlock it? How to disable it is right there in the manual—it just takes a minute. Does Phyllis have to hold her phone at arm's-length on speakerphone, shouting as she walks down the street? Does everyone at gate C12 need to hear Erik's video game? I love concerts. Rock concerts *should* be loud, but why not lower the volume on the house sound system between sets? It might be nice to use that time to discuss the performance with a friend without having to shout or simply to recharge a bit.

A hundred years ago, if we wanted to hear music, we would need to seek out a concert or, much more likely, play it ourselves. There was always an element of active engagement. We had to make time for music, and our time was rewarded with gratification—and by increased activity in the reward circuitry of the limbic system. Dopamine is released, positively reinforcing it, making it an activity we return to.[37] Now, music has moved from the foreground to the background, from a signal to a noise. Music is imposed on us in airports, in elevators, at the grocery store, and when put on hold on the phone. Instead of actively engaging with the music, we find ourselves treating it as one more source of noise to ignore, to grind our teeth to. When it simply joins the unwanted grating chorus of uninvited sounds, it is not tuning our brains through active engagement, not teaching us to pick up on the important details in sound, nor engaging our emotions fruitfully. We have learned to ignore it. How can that be good for our evolving sound mind?

Noise-Reduction Technology

A straightforward way to reduce noise is to use sound-isolating earplugs. Most are made of foam, and their one-size-fits-most design can work well, although I personally have difficulty with them because my ear canals are so twisty they fall out. I prefer the wax earplugs that mold to the contour of my ear canal and stay in better, especially at the gym or while I sleep in a noisy place. You can also look into custom-fitted earplugs. Often called "musician earplugs," these are designed to lower the sound level across the entire frequency range equally, so you don't selectively lose the high or low frequencies. Some custom-fitted earplugs have exchangeable filters to give you more or less sound reduction depending on the circumstance. For example, you might use an 8 dB filter while riding the subway but a 25 dB filter when playing drums. During the years-long construction period outside my office window, I wore my custom earplugs daily. This hugely reduced how much the noise bothered me.

Active noise-canceling headphones excel at mitigating continuous noise sources like airplane or train noise. They do so by generating out-of-phase sounds that are played simultaneously with the unwanted sound. The two opposing sounds cancel one another out but can also create more sound pressure. Some people, myself included, tend to feel fatigued after wearing these devices for a while. Both active and passive noise-attenuating earplugs have varieties with audio playback capabilities—you can listen to music, audio-books, or podcasts at the lower volume a reduced background noise level affords you.

When performing live, musicians often use a stage monitor, a speaker that delivers the sound of their own instrument toward them, so they can better hear what they are playing. In-ear monitoring solutions offer advantages worth considering. Well-mixed sounds can be fed directly (and wirelessly) to the ear from the soundboard. And the custom-fit ear mold attenuates the other sounds being produced on stage, the drums for example. Moreover, mobility is enhanced—a singer does not need to worry if she is not close to a stage monitor. Due to the noise reduction, she also will experience less vocal/motor fatigue because she will be less tempted to shout over the sounds of the instruments or the crowd. Finally, in-ear monitors minimize acoustical variations between venues.

Although not a noise in the sense we have been using here so far, reverberation (echoing) also interferes with understanding speech and distorts music. Using foam tiles, rugs, and tapestries diminishes reverberation that impedes understanding what we want to listen to. Architectural design for restaurants, music venues, and other public spaces is increasingly receptive to concerns about noise. Acoustic baffling is often found in orchestra pits and on restaurant ceilings. Microphones can pick up ambient sound and speakers play it back in such a way that reverberation is minimized—reminiscent of the approach in sound-canceling earphones. (On the flipside, some spaces use active acoustics to *enhance* reverberation to liven the acoustic environment.) In addition to sound-level measuring apps, there are crowd-sourced apps that rate public spaces by their noise-friendliness. Looking for a quiet place to study or a spot for quiet conversation? It is increasingly possible to patronize places where someone gave some thought to the acoustic design.

Hearing aids incorporate digital noise-reduction technology to differentiate between speech and noise on the fly.* The hearing aids can be programmed to enhance certain sound ingredients (your partner's voice) and cleverly apply amplification or attenuation instantaneously at appropriate frequencies to enhance speech sounds while ideally suppressing the clattering of dishes coming from the restaurant's kitchen. In this way, hearing aids become a tool to facilitate listening rather than just hearing.

It is true that many of the best options are also the highest in price. Noise-isolating custom ear molds cost a lot more than off-the-shelf earbuds. In-ear audio monitors are expensive. Noise-reducing hearing aids add cost to an already pricy health-care expenditure. Quieter hair dryers are out there, but they are double the cost of run-of-the-mill models. But until we get better as a society at caring about noise, products like these will remain niche and expensive. In the meantime, there are low- or no-cost things that we can do for ourselves and for our neighbors.

Attitudes

I have been to concerts where the musician boasts, "We're going to play so loud we'll make your ears bleed," and the audience responds, "Yeah!" There is a toughness that goes along with listening to loud sounds that can be destructive. This tough attitude is not unlike how we used to think about athletics—getting back in the game immediately after getting hit in the head. "Shake it off!" Consider seat belts and airbags in a car or safety protection in sports. As recently as the

---

*Noise is usually constant in its overall "shape" (frequency spectrum). In contrast, the acoustic properties of speech are more varied, but reasonably similar across talkers in syllable rate, dynamic range, and frequency content.

1970s, only a smattering of professional hockey players wore helmets and major league baseball players would shed theirs as soon as they reached base. Today, a helmetless hockey player seems inconceivable. Baseball players wear their helmets while on the basepaths and varieties with extended jaw guards have become the norm. We now appreciate the importance of protecting ourselves against concussions. Today, even most macho halfwits wear seatbelts, and more attention is being paid to safety in sports even to the extent that, for better or worse, contact sports are on the wane. It is my hope that we stop being so cavalier about noise in the same way.

Attitude changes are in the air. People like Gordon Hempton, with his Quiet Parks initiative, are working to preserve silent spaces.[38] People everywhere noticed and appreciated the reduced sound levels during the coronavirus shutdown. When noisy life resumed in Paris, noise complaints increased, especially about noisy motorbikes. Police anti-noise brigades stepped up their patrols and street-corner noise sensors were installed to issue automatic fines to motorbikes exceeding permitted noise levels.[39]

Our sound minds affect the choices we make in our sonic world. The less we appreciate silence and the more our brains become accustomed to noise, the noisier the world will get. A vicious cycle. Encouragingly, there are many opportunities to enrich the sound mind. Engagement with the right sounds can serve as an antidote to noise, as we saw in intensive-care babies, bilinguals, and musicians.

12

Aging and the Sound Mind

Don't shout. I can hear you, but I can't understand you.

My three sons are close in age and sometimes competitive. I would often leave notes in their school lunches. One note that would periodically make its appearance in all three lunches was "You are my favorite child." I think they were savvy enough to realize I didn't have a favorite, or they eventually figured it out by gloating about it and discovering they all got the same note.

Like my sons, I have no favorites among the thirty-odd doctoral students that have trained at Brainvolts to date. But one who would definitely warrant a lunchbox note was Samira Anderson. Unlike most who came to Brainvolts in their twenties, Samira was already a self-described "old lady" when she entered graduate school. She had worked as an audiologist in a mix of private practice and medical settings in Minnesota for thirty years.

Her clientele—mostly elderly people—informed her interests. Namely, what role does hearing play in the aging process and vice versa? Under her tutelage, Brainvolts began its investigations into the older sound mind. Samira was absolutely the best person to spearhead this line of work. As a clinician, she craved a biological understanding of the conditions she had spent decades treating. And her study participants loved her! She was generous with her knowledge of what happens with communication as we get older and shared it freely with the broad cross-section of people eager to work with her. For years after our "aging brain" projects wrapped up, some of her participants would call Brainvolts wanting to know if we were still looking for research subjects.

On one hand, we know a lot about the aging process as it pertains to the ear—that is, the cochlea. As we age, a combination of our cumulative lifetime of noise exposure and degeneration of the components of the middle and inner ear takes a toll on our hearing thresholds. Thresholds—the quietest sound levels we can hear—change in a characteristic way as we reach late middle age. One study found 46 percent of people older than age forty-eight had a hearing loss.[1] Another found hearing loss in 63 percent of those age seventy and up.[2] This ear-focused hearing loss is known as presbycusis—from Greek *presbys* + *akousis*, "old hearing." An experienced audiologist like Samira could probably look at a person's audiogram and guess their age within five years with no other information. But what Samira—indeed anyone—could not guess is how well that individual could *make sense of the sounds* they actually were able to hear. Indeed, even with normal hearing thresholds, some older people simply cannot understand the sounds they can hear. This failure to make sense of sound usually takes the form of difficulty understanding speech in noisy places. Figuring out why this is, and what we can do about it, has been a holy grail of hearing science.

In addition to age-related cochlear deterioration (generally affecting hearing in the higher frequencies), the hearing centers in the brain can also deteriorate. Sometimes this is a consequence of diminished input from the ear.[3] The brain needs sound to function optimally. Wearing hearing aids is accompanied by improved memory and listening in noise. Notable are the improvements in the brain's response to sound ingredients.[4] It is common to hear people say that wearing hearing aids helps them "think better." For that matter, I will put my contact lenses in before getting on a phone call; being able to see well helps me think.

Brain changes are often unrelated to hearing loss.[5] Indeed, aging is accompanied by a number of physiological changes in the sound mind. Aging can disorganize the tonotopic maps that distribute the processing of different frequencies and hinder the inhibitory processes that fine-tune frequency selectivity.[6] In addition, there is slowing of neural timing,[7] a reduction in connectivity among relevant brain regions,[8] and increased neural noise.[9]

Moving outside of the auditory system entirely, physiological changes happen throughout the brain with aging. As we get older, hemispheric activation can become more symmetrical, blood flow is reduced, and our brains can shrink—about 5 percent per decade after age forty, with both gray and white matter affected.[10] Mild cognitive decline involving processing speed and memory can occur.[11] System-wide changes in neural processing with age can further hamper the ability to make sense of sound.[12] I am talking about the day-to-day nuisances that start to become more common with age, like suddenly having difficulty calculating tips on a restaurant check or remembering what just happened in the novel you are reading. Cognitive difficulties with aging tend to involve this type of in-the-moment problem solving, which peaks somewhere in our twenties. In contrast, crystallized intelligence—the skills, knowledge, and ability learned and relearned over a lifetime—continues to *improve* into the seventies.[13]

And then there is dementia. It is not a specific disease; it is a group of symptoms that, along with memory loss, include confusion, decreased ability to concentrate, misplacing things, confusion about time or place, and often personality changes. Alzheimer's disease is the most common form of dementia, and it is estimated that 50 million people worldwide are living with it today. The extent to which dementia is directly correlated with any of the physiological brain changes listed above is not clear. Post-mortem examinations of people with and without Alzheimer's tend to be inconclusive; the degree of atrophy or degeneration often has little bearing on presence or severity of cognitive decline.[14]

What *is* clear is that sound is a door to our memory when other connections to the world are erased by dementia. Nancy Gustafson, a world-class opera singer, shares a personal anecdote about her mother, whose dementia had progressed to the point where she no longer recognized Nancy and was unable to speak more than single-word yes or no replies. One day, Nancy sat down at the piano in her mother's memory-care facility and began playing Christmas carols. Almost immediately, her mother began to sing along and was able to carry on a conversation for a while. Nancy founded Songs by Heart to encourage singing in people with dementia living in eldercare facilities. Music can address emotional and cognitive health in people with dementia.[15]

Signature of the Aging Hearing Brain

With all this in mind, an audiologist like Samira who is trying to help a seventysomething client make the most of his hearing must be conscious that his difficulties understanding speech might have at best only a cursory relationship with his ear-centric hearing thresholds. His difficulties may stem from changes aging has wrought on both auditory and nonauditory parts of his brain.

With Samira at the helm, Brainvolts began a large-scale project investigating the auditory brain signatures of older adults. Using the frequency following response, we asked just what the auditory-brain signature of aging is—what sound ingredients were affected? Then we looked at how consequences of aging on the sound mind might be slowed or reversed.

It is not very convincing to proclaim that a physiological response to sound in an "old" brain is, for example, smaller than a "young" brain if the former is suffering from presbycusis. If the sound is not being transmitted from the ear to the various auditory brain stations, we wouldn't expect the brain's response to that sound to appear normal. So we took a two-pronged approach to minimizing hearing thresholds as a confounding variable. First, we did our best to match audiograms. Although the statistics on hearing loss I cited above are somewhat grim, there certainly are sixty-to-seventy-five-year-olds with normal hearing thresholds. And we were able to rustle up some people with hearing loss in our "young" group as well to balance the scales. Second, we applied custom amplification to our sounds. We carefully recorded hearing thresholds across the frequency span on all our participants and, based on their unique profiles, crafted individualized sounds for everyone. Gene got a progressive boost just from 1,000 to 4,000 Hz, matching his sloping hearing loss, while Marjorie got a flat bump across the board. Thus, to the best of our ability, we ensured that each subject was listening to sounds that activated their ear equivalently.

Even with audibility equated, there was still a nearly across-the-board decline in the brain's response to sound in older listeners, as measured by the FFR.[16] There were some nuances, but generally speaking, in our older subjects, responses were smaller. They were delayed. They were less stable (consistent). They were less synchronized. The harmonic content was attenuated. (See figure 12.1.) Most striking was response timing, which makes sense because processing-speed declines that come with aging can stem from changes in white-matter integrity.[17] There was a delay in the brain's processing of speech syllables, particularly those with complex timing such as the FM sweeps contained in words like "dog." We saw delays in the response to these FM sweeps of as much as a millisecond or more—a lifetime in the hearing brain. The sound mind was simply not reacting as speedily as it had years before. What's more, there was a relationship between the degree of response degradation and how participants reported their experience. The participants who claimed their ability to hear in noise was basically OK had responses that showed less deterioration; those who reported having difficulty did, in fact, have a more deteriorated brain signal.[18] Remember, thanks to custom amplification, they were all "hearing" the same *ear* signal; so these self-reports suggest the *brain* signal we recorded represented the root of how well they were making sense of sound.

Can improving the sound itself reduce cognitive aging? Older people with hearing loss were given hearing aids for six months. Not only was their listening in noise and cognition better *even when their hearing aids were removed*, but their sound minds showed signs of reorganizing.[19]

We cannot determine whether the auditory brain regions were failing to respond adequately in their own right or whether they had become sluggish due to being starved of input from cognitive centers that had independently declined with age. Either way, we saw clear evidence of a problem in the aging sound mind that cannot be blamed on the ear alone. Even with the best hearing aid in the world prescribed, fitted, and programmed by a top-notch audiologist, the aging brain may struggle with tasks like pulling speech out of background noise.

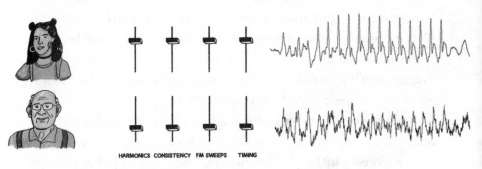

HARMONICS   CONSISTENCY   FM SWEEPS   TIMING

Figure 12.1
Older adults have a brain signature that shows declines in many facets.

What is the solution then? Samira was determined to figure out how she could help her older clients restore the sound mind that age had begun to take away.

Staving Off Auditory Aging

Training

Coinciding with the ubiquity of the personal computer and smartphone, there has been a flourishing of computer-based "brain training" apps. Some targeting older adults, others targeting school-age children, they tout an ability to improve memory, cognition, and attention by "rewiring the brain." Some have a plausible basis and are scientifically backed; others are probably just trend-chasing cash grabs. Among scientists, neuro- and otherwise, there is a mix of support and skepticism.[20] Nevertheless, Samira realized that now, armed with an objective way to gauge their effect on the sound mind, she could measure whether the older brain's response to sound could be strengthened by readily available training. If so, this would be important news for all of us and for hearing health providers.

Samira picked a commercial product that placed a special emphasis on *auditory* training. In particular, it included drills that directed attention toward particular sound ingredients, including distinguishing sounds (like FM sweeps), syllables, and words that differ in their timing. These are presented in increasingly complex listening contexts. The sound ingredients are easy to hear at first, but are modified to become increasing difficult based on the participants' performance as they learn to hear increasingly subtle sound nuances. It seemed ready-made to address the sorts of issues her aging clientele dealt with and the brain signature she had uncovered. She recruited seventy-nine people between the ages of fifty-five and seventy and randomly enrolled half in eight weeks of brain-training exercises and the other half to view and take tests on educational documentaries over the same time period. Regardless of group, the activity was performed for one hour per day, five days per week. Before and after the eight-week training regimen, everyone received tests of memory, listening in noise, processing speed, and FFRs.

After eight weeks, the people who were enrolled in the brain-training exercises showed improvements in memory, listening in noise ability, and processing speed. Their neural timing also sped up, especially in response to the FM sweeps of speech syllables presented in a noisy background.[21] None of those changes were seen in the people who viewed educational programming. It appears that directed auditory training over a relatively short time period can offer tuning of the sound mind and mitigate one of the chief complaints among older adults—that of difficulty navigating auditory scenes like hearing speech in noise. Samira recalls one of the brain-training participants, Fred, who couldn't believe how much better he was able to hear movies. "Suddenly I was laughing at the jokes and wasn't constantly wondering 'Who's that guy again?' It seems like sharpening my hearing sharpened my whole brain!" Another participant, Sandy, reported she was enjoying the noisy gatherings with her grandbabies more. Unfortunately, there are indications that these gains may not persist;[22] perhaps "booster" sessions are in order? Nevertheless, if chosen wisely, it seems that a brain-training exercise regimen could be a way to regain some of the timing precision that aging can compromise.

But what if there were a way to prevent this loss of timing in the first place?

Healthy Aging

With a growing aging population that is living longer and longer, there is increasing prominance given to the idea of healthy aging. The National Institute on Aging identifies four elements that can contribute to a productive and meaningful older life: maintaining a healthy weight, watching your diet, being physically active, and participating in hobbies and social activities. Doing these things has been linked to a lower risk of dementia and a longer life span.[23]

Conspicuously absent on the NIH list is the role of the sound mind in healthy aging. Yet the quality of life of an older adult is tightly linked to sound and hearing. Even with all other risk factors—age, sex, education, etc.—carefully controlled for, the existence of hearing loss is strongly and independently associated with cognitive impairment.[24] And among those with a dementia diagnosis, the rate of cognitive decline is accelerated in hearing-impaired individuals.[25] Both the NIH and the UK's counterpart have identified hearing loss as one of the most modifiable risk factors of dementia.[26] The dementia-hearing link is present in the hearing brain as much as it is in the ears. Listening-in-noise ability—which requires not only hearing the signal but the ability to *think* about it—is reduced in older adults with Alzheimer's disease and other forms of memory impairment.[27]

There is another pernicious angle to the hearing-dementia link. Hearing problems, whether in the form of decreased audibility over-all or difficulty hearing in noise in particular—is *isolating*. If you cannot hear speech well, you are less likely to go on outings with friends, attend church, call your children, or chat with the cashier at the grocery store. You tend to withdraw more and more, feel increasingly socially disconnected and lonely, and ultimately lead a less enriching life. These social factors—which *do* make the NIH list—are linked to dementia.

Just as a young adult can start exercising and eating well today to set herself up for healthy aging, there are things we can do for our sound minds that can pay dividends later on. Healthy aging begins in childhood.

Keeping the Sound Mind Young with Music

Musical training can contribute to a healthy older life. Listening to speech in noise is better in older musicians than in their nonmusician peers, and we can see this reflected in the brain's response to sound.[28] Moreover, older adults with musical experience maintain better memory and cognitive skills than nonmusicians.[29]

Brainvolts looked at auditory-brain function in older musicians. We recruited musicians and nonmusicians between the ages of forty-five and sixty-five. The musicians had decades of continuous musical practice, beginning in childhood. After carefully screening for normal hearing and IQ, and matching the groups on cognitive ability, physical, and social activity, we tested their hearing-in-noise ability. The musicians were better at listening in noise.[30] Then we looked at what effect making music had on the aging brain signature we had discovered previously. Remarkably, the declines in processing of all sound ingredients—timing, consistency, the works—were reduced or even nonexistent in older-adult musicians. Their brains' responses closely resembled those of healthy young adults (figure 12.2).[31] Even older adults who *do* have ear-based hearing loss benefit from music making—listening-in-noise ability in older adult musicians with hearing loss can match or exceed that of nonmusicians who have normal hearing thresholds, even those half their age.[32]

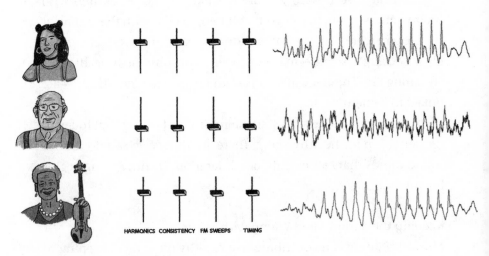

HARMONICS CONSISTENCY FM SWEEPS TIMING

Figure 12.2
An older musician's auditory brain appears similar to that of a young adult.

With or without hearing loss, the musician brain keeps producing crisp, young-adult-like neural activity into older age.

A little goes a long way   Late in life, arthritis robbed la mamma's hands of strength leaving her with swollen, painful knuckles. She was unable to open jars and it was difficult for her to tie her shoes. But, thanks to her auditory-motor memory born of a lifetime of music making, she still maintained her ability to play the piano.

The positive outcomes of making music can last *even if you do not continue to play music*. I often ask my audience, "How many of you have played music at one time in your life?" Many hands go up. "How many of you are still playing now?" Most of the hands go down. Many of us had some musical training in the past. Just like investing a modest sum of money at a young age can pay off handsomely at retirement age, we wondered whether playing some music earlier in life could pay off even decades later. Once the sound mind, through music making, has learned to effectively connect sound with meaning, does it continue to reinforce this skill automatically later in life?

Older adults with as few as three years' experience playing an instrument *many decades before* exhibited signs of a "younger" brain.[33] Specifically, there was more robust timing to acoustically demanding ingredients like FM sweeps in speech. This result aligns well with complementary findings in animals demonstrating that auditory enrichment early in development yields better auditory processing later in life.[34] The benefit, though, was more modest than in the older people who had kept up their music making. It was the lifelong musicians who showed enhancements in *all* of the sound processing ingredients shown in figure 12.2. Other investigations of early musical training found that older adults with at least ten years of musical training had better memory, executive function, and cognitive flexibility than an otherwise well-matched group with few or no years of music.[35]

It's never too late   What if you are older and you have never taken part in any music making? Will beginning musical activity *now* help?

Yes! As we saw with the eye-prismed barn owls and other animals,[36] the human sound mind continues to be shaped into old age. An older person who begins to make music *today* can see benefits in both neural processing and real-life listening abilities. Ten weeks of two-hour weekly group choir sessions along with weekly vocal-training sessions was associated with an improvement in listening in noise and boosted neural responses to the fundamental frequency in speech (a voice pitch cue) in adults ranging in age from the mid-fifties to late seventies.[37] Learning piano later in life led to better listening in noise and a strengthening of the brain's speech-motor system.[38] Another study pitting music listening against music playing found that the sixty-to-eighty-year-olds who actually made music improved their working memory and hand coordination.[39]

Inspired by older adults' pervasive participation in singing groups in Finland, University of California professor Julene Johnson embarked on a large-scale study. She found a decrease in loneliness and an increase in quality of life among older adults who participated in community choirs.[40] Quantifiable health outcomes like numbers of doctor visits, prescription medications, and falls were lowest in older adults participating in a choral group.[41] So making music has a direct impact on the sound mind in older age, in addition to other benefits that music making can bring: improved quality of life, sharper memory, and an overall boost to well-being.[42]

Keeping the Sound Mind Young by Speaking Another Language

Sources of cognitive health include cognitive exercise, education level, diet, physical activity, and an active social life. Bilingualism is another factor to add to the list. Bilinguals typically outperform monolinguals on tasks that require cognitive skills like attention and inhibitory control. This edge is maintained in older bilinguals.[43] In people with Alzheimer's disease, the bilingual brain is able to tolerate a greater extent of brain degeneration before performance suffers.[44] Some studies have attempted to pin a number on this finding and have claimed that speaking a second language can delay the onset of dementia four to five years.[45]

Embracing Aging

To tell the truth, I enjoy getting older. Someone my age has simply had more time and material to work with than a teenager. My life experiences—the sounds I have loved and lived with over the years—have made me me: the sounds from underneath my mother's piano, the sounds of the mountains in Italy, the sounds of New York City, the sounds of my electric guitar in my twenties, the sounds of my favorite sons' voices, the sounds of my electric guitar in my sixties, the rock opera I will write when I'm ninety . . . my own sound mind promises to evolve.

I have attended conferences on aging where the prevailing message is OLD = BAD. This conclusion comes from looking at factors we can measure: hearing thresholds, reaction time, atrophy in the brain. What's missing is research on the immeasurable: wisdom, patience, compassion, joy. With age we learn how to listen and what is worth listening to. The product of life experience is impossible to measure, but if it *were* measurable, I think there would be more conferences with the theme OLD = AWESOME. (But maybe this is just a distorted perspective coming from my cognitively enfeebled mind.)

I hope the link between hearing, thinking, and how we feel will become increasingly recognized. Time travel is still out of reach, limiting what could have been done years ago to build up our sound mind. But learning (or relearning) to play music, studying another language, and training that strengthens sound-to-meaning connections offer possibilities. Our sound minds are a conduit to living a richly interconnected life.

13

Sound and Brain Health: Spotlight on Athletes and Concussion

There's more than one way to tune . . . or hurt . . . the sound mind.

My uncle Hans was an orthopedic surgeon, a skier, and a rock climber who made dozens of first ascents in New York's Shawangunk Mountains and in the Dolomites. Hans was an advocate of physical fitness in children, and his research had a lasting effect on school curricula.

In the 1950s, Hans championed the idea that all children should get mandatory physical education in school, and that fitness training should not be reserved only for those who aspired to play varsity sports. He took this position after studies found that American children were less physically fit than their European counterparts.[1] After administering the six-part flexibility and strength evaluation called the Kraus-Weber Fitness Test to thousands of children in the US, Austria, Italy, and Switzerland, Uncle Hans uncovered some sobering statistics: 58 percent of US children failed at least one of the six measures while only 9 percent of the European children did.[2]

Hans reported his findings to President Eisenhower, who formed the President's Council on Sports, Fitness, and Nutrition. The later

1950s and 1960s saw a tremendous growth in physical fitness programs in the public schools. Hans's point of view about youth fitness resonates with my view of music education. Neither fitness
nor music training should be meted out only to those students who
excel in these activities. Every child benefits from being physically
fit. Fitness, like music, should be an integral part of every child's
upbringing.

Remarkably, Hans's stance on youth fitness was something of an
outlier at the time. Today, we know athletic training is one of the
best things you can do for your body. It promotes physical fitness,
improves cardiovascular function, sharpens cognitive skills, and
boosts neurological health.[3]

What do rock climbing and physical education have to do with
the sound mind?

An Upside of Sports: The Athlete's Sound Mind

The brain is affected by athletic activity. Learning a new physical
activity as an adult can increase the brain's gray matter volume
and sharpen cognitive skills.[4] A direct link exists between myelin, a
neural insulator that increases the speed of communication among
neurons, and learning a new skill.[5]

Lurking inconspicuously in the background of the body systems
strengthened by athletic participation is the hearing brain. An athlete will tell you that sound plays a role in her performance, from the
obvious (listening and responding quickly to teammates' cues and
coaches' instructions) to the subtle (monitoring the sounds of activity
on the field to adapt one's own movements).[6] Athletes must rely on a
responsive and precise sound mind. Thus, Brainvolts asked whether
this can be corroborated physiologically in the brain's response to
sound.

We measured the brain's response to sound in nearly 500 Division I NCAA athletes at Northwestern University and in another 500 nonathlete undergraduates.

We looked at how large the response to sound was relative to the background neural noise that is always present in the nervous system. As with the static in the background of a radio signal, you can improve the situation by minimizing the static or boosting the announcer's voice. In our athletes and nonathletes, we computed how much bigger the response to speech is than the background noise. The sound-to-noise ratio was larger in the athletes. This was accomplished not by turning up the signal but by turning down the noise (figure 13.1).[7] This suggests that physical activity can drive "cleaner" sound processing in the brain, potentially enhancing communication.

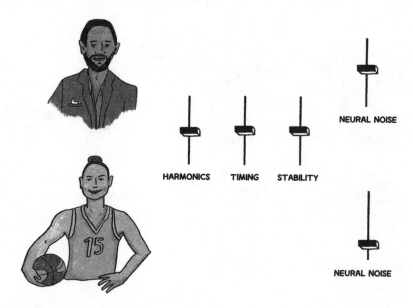

Figure 13.1
The signature of the athlete brain is a quieter brain. Sounds are enhanced because neural noise is reduced.

Like athletes, musicians and bilinguals have enhanced sound minds, but unlike athletes, they hear the announcer better by boosting his voice. Musicians show precise processing of sound ingredients essential for conveying which words are spoken (timing, harmonics, FM sweeps). Bilinguals have strong responses to the fundamental frequency, which helps lock onto the voice of the talker. The sound stands out to the athlete because it is unencumbered by background neural noise. Everyone hears the announcer well, but athletes, bilinguals, and musicians differ in how their sound minds make it happen (figure 13.2).

Background neural activity reflects brain health, as seen by differences in background noise across socioeconomic strata and increases in neural noise with aging and after acoustic trauma.[8] Neural noise is greater in the "linguistically deprived" brain. Reduction in the cumulative lifetime of linguistically rich sound-to-meaning

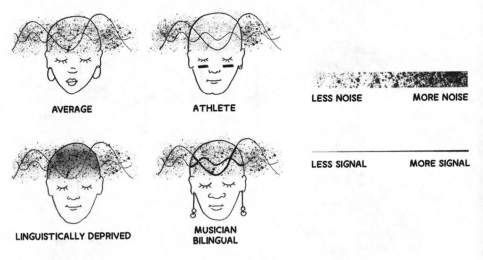

AVERAGE   ATHLETE

LESS NOISE   MORE NOISE

LESS SIGNAL   MORE SIGNAL

LINGUISTICALLY DEPRIVED   MUSICIAN BILINGUAL

Figure 13.2
Compared to the average listener (top left), linguistic deprivation (bottom left) both turns up the noise and turns down the signal, making the signal more difficult to hear. The musician's and bilingual's brains (bottom right) turn up the signal while the athlete's brain (top right) turns down the noise. Both strategies make the signal easier to hear.

connections leaves an excessively noisy brain poorly equipped to home in on important sounds. The athletes' brains show the exact opposite. The reduction in ongoing neural activity in athletes suggests a less noisy neural infrastructure for them relative to nonathletes, enabling crisper processing of sound. A quieter brain can work more efficiently to make sense of sound because of the reciprocal connections the sound mind shares with cognitive, sensory, motor, and emotion systems.[9] It remains to be established whether this mode of enhancement is tied to the overall fitness level of athletes, the heightened need of an athlete to engage with and respond to sound, or both.

A Downside of Sports: Concussion

Making sense of sound is one of the hardest jobs we ask our brain to do. It stands to reason that getting hit on the head can disrupt this delicate and precise process. Sound processing in the brain can provide biological insight into concussion.

The prevalence of concussions, also known as mild traumatic brain injury (mTBI), in sports is in the spotlight. Contact sports are among the most popular. Americans love football; Super Bowl viewership routinely doubles that of its closest competitors, typically a presidential address or debate.

From 2012 to 2019, there were on average 242 diagnosed concussions annually in the National Football League. This is a 7 percent concussion rate. The list of players in the NFL retiring following a concussion is long. Soccer, rugby, hockey, and other sports are facing an increase in early retirement. Well-known former NFL players, including Tony Dorsett and Jim McMahon, are suing the league for failing to adequately inform players about the link between concussions and long-term health problems.

Another estimate of concussion prevalence in the US, found that over 65 percent of the 200,000-per-year sports-related emergency room visits for head injury were in children under the age of eighteen.[10] Some prominent former NFL players have called for eliminating tackle football before the age of fourteen.[11] Calls for schools to modify play in contact sports are no longer considered fringe opinions.[12]

With the threat of concussion and repeated blows to the head, there is risk of both short-term and long-term brain damage to contact-sport participants. In particular, a condition called chronic traumatic encephalopathy (CTE) has been identified in dozens of retired NFL players postmortem. CTE is defined by cognitive impairments including memory, processing speed, and decision-making. Coined around 1940, CTE describes a condition previously referred to as punch-drunk and dementia pugilistica. The author of a 1928 *Journal of the American Medical Association* article noted that in punch-drunk boxers "marked mental deterioration may set in necessitating commitment to an asylum."[13] Indeed, anger, depression, impulsivity, and other changes in mood are common long-term results of repeated head injury sometimes noted in ex-NFL players.[14] Some have been diagnosed with CTE postmortem, too often following suicide.* Participants of contact sports such as football or boxing can also experience "subconcussive" injuries. These injuries are not severe enough to cause acute concussion symptoms, but the accrual of subconcussive events over time is believed to lead to progressive brain atrophy and CTE.

The prevalence of CTE is difficult to ascertain, in part because people who have their brains evaluated for CTE more often than not are those whose histories of repetitive brain trauma and troubling behavior suggest a high probability of a CTE diagnosis.[15] Acknowledging this built-in bias, a *JAMA* report in 2017 found that of 111 ex-NFL

---

*Currently, a postmortem examination of brain tissue is the only way to definitively diagnose CTE. Evidence of CTE hinges on the presence of abnormal bundles of phosphorylated tau (p-tau) proteins in the brain, a characteristic CTE shares with Alzheimer's disease.

players who had their brains examined after death, 110 showed some evidence of CTE, with 86 percent considered "severe."[16]

Sports organizations at all levels are reevaluating practices to prevent or minimize head injuries.* Along with such rule changes comes the necessity for timely, accurate, and portable assessment of head injury. Objective biomarkers that do not require the athlete to actively participate in testing have obvious advantages. After a head injury is not the best time to ask an athlete to actively perform a test. Moreover, there is a culture of perseverance and team loyalty among athletes that can result in the underreporting or masking of symptoms. An athlete having sustained a head injury might try to shake it off, insisting she feels fine and can return to play. Moreover, an athlete might try to game the system by deliberately performing poorly in baseline testing. For example, one such test asks an athlete to stand on one leg for a set amount of time. However, a wide receiver who knows he is likely to receive some big hits in a football game may intentionally feign a bit of wobbliness at the preseason baseline testing. "See coach, I couldn't balance so well before my hit either. I'm fine to go in for the next play." Ideally, there would be a measure where "the brain does the talking."

---

*For example, after a study noted that more than 20 percent of football concussions occur on kickoff plays, the Ivy League adjusted the kickoff position from the 35 to the 40 yard line, resulting in doubling the rate of touchbacks and a corresponding reduction in high-speed, go-for-broke return attempts. Concussion rates dropped in the first year of the new rule. In the NFL, since 2016, touchbacks are now placed at the 25-yard line to better incentivize a receiver not to risk a return when the ball is caught in the end zone. There are even reports that the NFL is considering eliminating kickoffs. Tougher officiating and harsher penalties have reduced the number of illegal blocks and instances of roughing the passer or the kicker. Independent physicians monitoring the game are authorized to remove players from the game to go through concussion protocol. In other sports, the European soccer federation is pushing to increase the time allocated for head-injury assessments. The height of an illegal "high tackle" in World Rugby is being lowered in a bid to reduce the incidence of head injury.

There are good reasons to think the auditory system might be a fruitful avenue to reduce the ambiguity inherent in concussion diagnosis. We now understand some of the sensory, cognitive, motor, and emotional repercussions of concussion.[17] Each of these brain systems is entwined with the sound mind.

Brain Assessment with Sound: A Brief History

There is precedent for using sound in the diagnosis and management of brain injury and other neurological conditions. Neurologist Arne Starr (whose watercolor *Neuralscapes*[18] were reproduced in some of the figures in chapter 2) pioneered the use of auditory responses, measured with scalp electrodes, as an index of neurological health. The subcortical auditory system is the brain's timing expert. Brain tumors, strokes, multiple sclerosis, and other neurological disorders can have a harmful effect on the *timing* of neural responses to sound.

When everything is working as it should, the hearing brain is a marvel of timing precision; it is only due to this precision that synchronized responses in deep-brain structures manage to eke their way to the surface of the scalp as recordable electrical fluctuations. The tiniest amount of disorganization in timing is enough to squelch or delay this minute signal, or to prevent it from ever seeing the light of day. A neural peak or trough occurring even a fraction of a millisecond later than expected gives us a strong indication that there is something worrisome going on in the brain.

Historically, tests of subcortical timing were limited to responses to sound onset. But now, it is possible to see how the brain processes other sound ingredients (pitch, timing, timbre . . .) using the frequency following response (FFR). Diagnostic uses continue to emerge, including disparate conditions that cannot be visualized with MRI, such as schizophrenia, ADHD, autism, language impairment, hyper-bilirubinemia, and HIV.[19]

Concussion and the Hearing Brain

No single test can diagnose a concussion. Even if MRI were fast, cost-effective, and portable, you rarely see evidence of a concussion with brain imaging. A physician must weigh the results of a variety of tests along with a patient's sometimes-unreliable reporting of symptoms. Furthermore, symptoms and impaired cognitive performance can be transient or may not emerge immediately after contact. There are guidelines for diagnosis, such as observing symptoms spanning somatic, cognitive, emotional, behavioral, or sleep domains, which cannot be explained by a preexisting condition, medications, or drug use.[20] Yet two physicians assessing the same patient may quite reasonably come to different conclusions. And the stakes are high. If the individual making the assessment is an athletic trainer on the sidelines of a football game determining whether an offensive tackle should return to the line for the next snap, the wrong decision could endanger an athlete or negatively impact the outcome of a crucial game.

Much of what we know about a concussion's effect on sound processing comes from observations of soldiers who suffer traumatic brain injury or concussion from roadside bombs and other circumstances where a blast is involved. Unsurprisingly, if you are close enough to an IED (improvised explosive device) to receive a concussion from the blast, the sound of the blast is liable to injure your ears. For a long time, there was no particular reason to think the physical impact of the explosion—the brain injury itself—caused any resulting sound processing difficulties. Rather, it was assumed that any auditory problems that followed were simply due to damaging sound exposure. But evidence is mounting that "silent" head injuries also might have a detrimental effect on making sense of sound.

People who have sustained a concussion can have difficulty performing auditory tasks. These tests range from tone-pattern recognition (beep-beep-boop, "What did you hear?" "High-high-low") to speech perception. The most common complaint is difficulty hearing speech in background noise. Controlling for hearing loss, Erick Gallun looked at soldiers who had received traumatic brain injury but had normal hearing thresholds. He found their ability to hear speech in noise was three times worse than in matched control subjects.[21] Similar findings on *sports-related* concussion have emerged; athletes who have received one or more concussions in the past have difficulty processing sound.[22] Another indirect line of evidence that sound is involved in concussion is that auditory rhythm-based therapy is a promising line of rehabilitation in the recovery of cognitive skills after a concussion.[23]

Concussion often causes swelling, which compresses brain tissue.[24] Concussion can cause shearing or tearing of nerve fibers.[25] Some of the longest fibers in the nervous system connect subcortical and cortical regions of the brain. Nerve fiber integrity in the midbrain is reduced following collegiate football participation.[26] Concussion can disrupt auditory cortex function.[27] Subcortical timing to sound onsets is affected by head injury,[28] with a correlation between the severity of the injury and the extent of the timing delay.[29]

Concussion in Child Athletes

Most concussed patients recover within a week, but in about a third of cases, symptoms persist for a month or more. Brainvolts partnered with Cynthia LaBella, a pediatrician and concussion specialist, to investigate auditory processing in these persistent cases. Cynthia directs sports medicine at a major children's hospital that sees about 300 concussion cases a year, most resulting from athletic injuries.

We tested children in her clinic who had persistent symptoms and were *actively symptomatic* after having sustained a concussion. Sure enough, these children had excessive difficulty hearing speech sentences in noise.[30]

This study gave us evidence of auditory processing difficulty with a concussion, even when the ear is normal and you can take hearing loss out of the equation. Motivated by the need for improved concussion assessment, we began to look into whether the damage that causes auditory processing problems following a concussion can be measured physiologically.

Dr. LaBella's sports medicine clinic sees children with musculoskeletal injuries (for example, sprained ankles, broken arms) as well as concussions. Brainvolts alumna Ellie Thompson tested these children on their hearing-in-noise abilities and obtained their FFRs. We discovered that *timing* and the size of the *fundamental frequency* identifies concussed children at an impressively high rate, while clearing controls (children with musculoskeletal injuries)* at an even higher rate.[31] What's more, these children, who were at various stages of the recovery process, had fundamental frequency responses that correlated with the severity of their symptoms, positioning the hearing brain to monitor recovery. Indeed, in the children tested at a second time point as their concussion symptoms were clearing, their auditory brain activity returned to normal. Subsequent work continues to solidify the sound mind-concussion link.[32] The fundamental frequency finding fits the problems in understanding speech in noise we had noted in concussed youth. We rely on voice pitch to understand speech in noise. Locking on to the pitch of a talker enables one

---

*In any diagnostic tool, one must tread a thin line between sensitivity and specificity. Sensitivity is the true-positive rate: how many concussions did FFR flag among those who were already otherwise diagnosed? Specificity is the true-negative rate: how many non-concussion controls were correctly cleared by FFR.

to treat their voice as a unified auditory object, helping it stand out from the distracting noises in the background.[33]

Concussion in College Athletes

Tory Lindley, assistant director of athletics and head athletic trainer, naturally wants Northwestern University to win. This includes being a leader in athletic health and safety.

I think of myself as an athlete—I regularly enjoy calisthenics, boxing, hip hop dancing, and bicycling. Way back when, I rode 3,000 miles cross-country in thirty-three days. But I do not follow team sports. Jen Krizman follows them all. Jen (who may be even more competitive than Tory) was tuned into the ongoing concern about head injuries in sports, and was instrumental in connecting the dots between auditory processing and concussion. Jen spoke the language of sports that I decidedly did not. She helped guide Brainvolts and Northwestern University Athletics into a partnership to look at sports participation holistically—the good as well as the bad—as it pertains to the sound mind.

In testing the football team, we began with twenty-five symptom-free athletes who had had one or more concussions in the past but were recovered by the time of their testing. Would their auditory brains reveal a legacy of their past injuries? We examined the brain's response to sound in these recovered-from-concussion athletes and compared them to twenty-five football players who were position-matched yet had never had a concussion. Like the symptomatic children we studied, the football players with a history of concussion exhibited reduced responses to the fundamental frequency.[34] Not only does the sound mind show promise in the assessment of active concussions, but it appears to be sensitive to *past* head injuries as well. Perhaps this work can contribute to the quest for early identification of CTE, which at present is diagnosable only at autopsy.

We have expanded our investigation of the sound mind and concussion to all athletes, male and female, across the population of Division

I athletes at Northwestern University. We test all five hundred of them at the start and end of every season. If an athlete sustains a concussion, we immediately evaluate the athlete and follow him at weekly intervals. We compare a given athlete's response to sound to his own baseline neural signature.

Pitch, timing, and harmonics seem to arrange themselves systematically with the stage of head injury. In the acute stage, all three ingredients show signs of disruption. The disruption of harmonic processing is the first to resolve as symptoms begin to clear. After recovery, timing is restored, although a hint of pitch encoding difficulty can leave a lasting legacy on the brain (figure 13.3).

As it progresses, our longitudinal study will inform us about the possible risk of playing collision sports even in athletes who escape a concussion. Because of its sensitivity, granularity, and vulnerability, the FFR is a strong contender to pick up on the subtleties of aggregate disruption in auditory processing that might result from cumulative

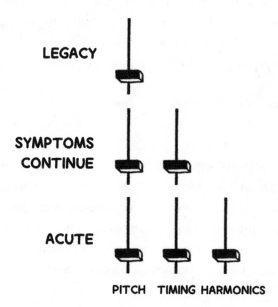

Figure 13.3
Stages of disruption of neural processing of sound following concussion.

*sub*concussive head impacts. Can four years of participation in a contact sport in the absence of a clinically recognizable concussion harm the brain? Or do we see only the athlete's "quiet brain" enhancement?

Returning to Play

"Does Beth need to come out of the game?" "When will Stu be cleared to play?" The chance of sustaining a second concussion increases following the first.[35] This is likely because the brain has not fully recovered, making the athlete at increased risk for future injury. Hopefully, a measure of the brain's response to sound will help determine when an athlete is ready to return to play.

Returning to Learn

The sound mind works in detecting brain injury *because auditory processing is compromised.* A young person who has recently had a head injury might not do well in a noisy classroom. This has implications when a child with a sports-related concussion returns to school. Clinicians and teachers are gradually becoming aware of how a damaged ability to make sense of sound might impact life outside the playing field in the classroom or workplace.

Vision, Balance, *and Hearing*

Vision and balance are routinely assessed following a concussion. What about hearing? With Dr. LaBella, team physician of the North Side Youth Football League in Chicago, we tracked neurosensory performance in young tackle football players across two consecutive seasons. Notably, each measure—vision, balance, hearing—contributed a distinct insight into brain health.[36] Performance on one of these domains could not predict performance on the other two, supporting their combined use in concussion assessment.

Summing Up

Sound rarely makes the news, especially in international politics. An exception was in 2016, when US and Canadian diplomatic corps stationed in Cuba reported hearing persistent, focused sounds. Upon examination, many of the diplomats exhibited typical signs of a concussion, including headaches and dizziness. In a news piece on the attacks, the *New York Times* referred to the diplomats' condition as "immaculate concussions." The source of the sounds remains a mystery, with suggestions ranging from targeted microwave blasts to lovelorn crickets. But whatever the sound source, a thorough assessment of the sound mind could help determine whether this kind of injury is in fact analogous to a concussion.

Building on the history of using sound to assess brain injury and other neurological conditions, an assessment of the health of the sound mind adds a new level of precision and potential. Incorporating auditory testing into standard practice of concussion management could improve health outcomes for athletes. Our growing knowledge of how concussions can affect the hearing brain promises to give us greater understanding of the sound mind in all of its complexity.

Physical training has positive effects on sound processing and contributes to overall brain health. I think Uncle Hans, who was happiest when he was clinging to a sheer rock wall, would agree. No matter what the activity—athletes need to train just as musicians do. My hope is to see physical health become a bigger educational and social priority.

14

Our Sonic Past, Present, and Future

Making sound choices for our sonic future

Sound Is Everywhere—Even Where We Least Expect It

Sound is a powerful force that shapes our sound minds and the world we live in. But so far, I have only touched on a part of sound's reach.

Plants can hear! We all know someone who talks or sings to his plants to entice them to grow. Indeed, scientists have looked at the validity of claims of sound affecting plant growth. In one case, the germination and seedling growth of jack pines was observed to accelerate when subjected to ultrasonic (too high-pitch for human ears) sounds.[1] In another, vibrations in the human-audible range (50 Hz) promoted seed germination and root elongation in rice and cucumber plants.[2]*

Any plumber can tell you that plant roots like to make their way into water pipes underground. Monica Gagliano looked deeper into

---

*Other plants discovered by scientists to be responsive to sound in some way include okra, zucchini, cabbage, chrysanthemum, pepper, and tomato.

this phenomenon by growing pea plants in forked pots where roots could grow to the left or to the right. Playing an audio recording of water at one branch of the fork (crucially, without water actually being present) caused the plant to reliably send its roots in the direction of the sound.[3] Moreover, plants, like vertebrate neurons, are tuned to particular sound frequencies. Corn roots in water will bend only toward the source of a 220 Hz sound and not to other frequencies.[4]

Plants use sound to gather information about their environment interpreting cues that are beneficial for their survival. A process known as buzz pollination happens when plants—including egg-plants, blueberries, and cranberries—release their pollen only when certain bees buzz at the correct frequency, somewhere in the 200 to 400 Hz range.[5] This prevents the "wrong" kind of insects—those without fuzzy bodies tailor-made for spreading the pollen around— from getting to the pollen.

Bioacoustics, the science of the relationship between animals and their acoustic environments, can be used to investigate sound produc-tion and perception. Ranging from underwater sounds—like whale songs that can travel for hundreds of miles—to echolocation in bats to birdsong, bioacoustics is a growing field.

The power of underwater sound and its deployment has helped in recovery of coral reefs. Coral reefs, naturally noisy places—from the clicks of seahorses to the grunts, purrs, or even barks of fish— produce a rich soundscape. When reefs begin to die, due to an extreme heatwave and overfishing, sounds diminish as their inhab-itants move away. Fewer residents mean less sound, making the reef less attractive to newcomers who use sound to judge its desirability for habitation. In an investigation to test the importance of sound, several new reefs were built in an area devastated by coral mortal-ity. Some of the new reefs were rigged with speakers piping in the sounds of a healthy reef, while in others, there was no sound. The sound-rich reefs attracted double the fish and other marine life as the silent ones.[6]

Another unexpected presence of sound comes in the airline meals served to us. Did you ever wonder why airline food tastes a bit off? A little blander than it ought to? Or why a disproportionate number of fliers seem to order tomato juice or Bloody Marys? Is it the dry air? The low pressure? The altitude? It turns out the biggest reason is sound. Loud noises like those of jet engines affect our taste perception. In particular, salty and sweet flavors are suppressed.[7] On the other hand, umami—a dominant flavor in tomatoes—is largely unaffected.[8] It may be that we gravitate to tomato juice because it is one of the few things that tastes "right" to us at 35,000 feet, while we can be left unsatisfied by the foods and drinks calibrated to a "quiet-appropriate" level of salty or sweet. From an evolutionary standpoint it makes sense that loud sounds would suppress your appetite. Who is hungry when there's an avalanche coming?

For better or worse, sound can be used as a weapon. Classical music has been used outside shops to dissuade teenaged loiterers. The US military has developed bona fide sound weapons, like a focused beam of sound that can be "shot" at individuals or groups (e.g., protestors that authorities want to disperse). The force of the wave can temporarily debilitate an individual from hundreds of meters away. There are narrow-beam sound technologies that can deliver sound at great distances to pinpoint locations—for example, to speak a warning to a distant unidentified boat that approaches too close to a naval ship. And there is still the open possibility that sound weapons were the source of the concussion-like symptoms that beset diplomats in Cuba.

A Word about Metaphors

The brain-as-computer metaphor remains unpersuasive to me. There is plenty we do not know about the brain and that includes the sound mind. But what we *do* know about the brain is that it works nothing like a computer.[9] In this book, I've made liberal use of metaphors, especially that of the sound mind as a mixing board. This metaphor, like any, has limits. In the case of the mixing board, it is inanimate, while the sound mind is alive and exists in the lived world. Like the second grader who uses hungry crocodiles to help him keep his inequalities straight, or the beginning electronics student whose understanding of the invisible flow of electrons is helped along with imagery of water tanks and pipes, the mixing board metaphor serves as a tangible expression of an underlying truth. Still, the neural processes it helps us imagine elude our full understanding.

Sound Connects Us to the Lived World

The other day I was walking outside in the town of Evanston, where I live. I was on the phone with my son who lives a thousand miles away when all of a sudden he interrupted the story he was telling me and exclaimed, "Evanston birdies!" He knows the sound of home. We all do.

We viscerally respond to the sounds of home—the neighborhood birds, the sounds of leaves rustling, the distant church bell, the abrupt hiss-honk of the city bus's air brakes and the pick-up basketball game down the street. Even the sound of the traffic when filtered by nearby houses and trees acquires a unique timbre as it arrives at my back porch. These all impart a sense of place, a place of belonging.

Over the years, I've learned public speaking goes best when I feel I'm speaking directly to the audience—no script, no reading, no podium. While my topics are thoroughly prepared, I allow room for spontaneity. I never know which words I'll use; I find them in the moment.* I have the same preference for making music. I'm happi-

---

*Did you ever wonder why some lectures are called "keynote" addresses? They set the tone of the proceedings, like establishing the key in a piece of music.

est with a deep knowledge of the structure of the piece, but I crave the room to improvise—no score between me, the sounds I'm making, and whoever may be listening. I let the music take me where it wants to go, but always with an eye toward bringing it back to the tonic. Back home.

Sound allows us to connect perhaps more than any other sense, even at a distance. Some believe the origins of music lie in mothers singing to their babies to establish bonds—so the infant could be comforted by her presence even if she were a little ways away attending to something—and then more broadly, to form cohesion in the larger social group.[10] Singing was the first music and music remains a strong social connector.

One of the languages I learned to speak is harmony. You simultaneously hear yourself and what your partner is singing, and use this feedback to adjust your own movements accordingly. This interaction is about modulating the space between voices, a marriage, a sensitivity to the other person and to the space between them. Singing harmony is emblematic of sound's power to connect us.

Sound is alive, created and experienced in the lived world.

Context and the Sound Mind

One evening at dinner with my son's girlfriend's family, I thanked her father for his wonderful daughter. "She is mostly self-made," he said. This, in a human-relations analogy, encapsulates much of what I have come to appreciate about the sound mind. The role of the ears—like the role of parents—is undisputedly crucial. And what our sound minds *do* with the sounds we encounter throughout our lives makes us who we are, sonically speaking. The sound mind brings context to the sounds our ears deliver to it.

We may have learned from our piano teacher that a B-flat is a perfect fourth above F. If we are geeky enough, we might know that the B-flat below middle-C has a fundamental frequency of 233 Hz. Somewhere along the way we may have learned the satisfying etymological origin of "malaria" is *mal aria*, "bad air." By themselves, these bits of "just the facts, ma'am" knowledge are meaningless without putting them in context. Being able to incorporate the knowledge of musical intervals into a musical composition or the knowledge of individual words into a novel provides that context. It is our sound mind's role to put the sounds it encounters into the context of our lives.

The sound mind affects the music we compose. Why didn't Bach use dissonance, meter, and rhythm in ways we have later come to know as jazz? Bach had the same twelve notes at his disposal. But Bach worked within the confines of his own sound mind shaped by the sonic environment he inhabited, just like anybody else.

As a group, bilinguals, musicians, dyslexics, and aging adults have distinct sound-mind signatures, but understanding *individual* sound minds is where the fascinating question lies. In *How Musical Is Man?*[11] John Blacking notes: "Everyone disagrees hotly and stakes his academic reputation on what Mozart really meant in this or that bar of one of his symphonies, concertos or quartets. If we knew exactly what went on inside Mozart's mind when he wrote them, there could be only *one* explanation." Mozart had his own sound mind. *Everyone* has their own unique sound mind.

Consider a sound—the "mighty da" is as good as any. Within this short utterance, there are timing cues and pitch cues. There is harmonicity. There are FM sweeps and a particular pattern of harmonic bands. We can measure, on a micro level, the brain's response to any of these sound ingredients. In isolation, it can tell us that someone's timing is a bit off or they lock on to voice pitch especially robustly. Or we can look at these attributes in the context of a person whose life experience has molded the hearing brain into a unified sound mind that processes all of the attributes as a whole. We can think of putting these facts together as something like filling in a survey:

Timing = □ early □ normal ⊠ late

Fundamental frequency = ⊠ big □ normal □ small

Response consistency = ⊠ consistent □ inconsistent

Etc.

Looking at only one attribute, we can say, "You have late timing" or "You have a consistent response." Alternatively, we can look at the profile as a whole and say, "You're a bilingual dyslexic!" or "You're Joey!" (figure 14.1). Sound and the sound mind's electrical response to it must be taken in context.

The hearing brain is a beautiful integrated system. But our auditory equipment on its own, though intricate and impressive, doesn't function in isolation. We would never learn anything about sound without relying on the context provided by our thinking, sensing, moving, and feeling brains to give meaning to the process of hearing.

Figure 14.1
The sound mind adjusts our sound processing, emphasizing and deemphasizing sound ingredients based on a lifetime of hearing, feeling, moving, and thinking with sound.

When we hear a sound, it is instantly accompanied by all its associated feelings, visual cues, and what we know about it ("That's an Italian accent"). We register these components *all at once* in what is called "perceptual binding." Scientists and philosophers have long grappled with how and where all the components we perceive come together. What we know about sound, how we feel about it, and what sights accompany it, influence how we make sense of sound, bringing us closer to an understanding of how it all comes together.

Our Sonic Personality

I am fascinated by the biological adaptations that take place over the course of our lives and unconsciously make us hear in our own way. I recently revisited Bach's Italian Concerto decades after I learned it on the piano. At first I was lost, then it started to come together, slowly at first, and then faster. Coaxing a memory out of my sound mind uncovered a part of me I didn't know was there. My conscious mind couldn't help me, but eventually the piece began to emerge.

Experiences like this have encouraged me to pay more attention to my gut feelings. We are cautioned to look before we leap, to weigh the pros and cons of every situation, to be rational. But if our gut is telling us something, maybe we should listen. This is because gut feelings are not arbitrary. They come from years of experience. Gerd Gigerenzer, in his book *Gut Feelings*, talks about how some situations resist reasoned calculations.[12] There is not enough information to be 100 percent sure of the right stock to invest in. You can look at past performance, study the financial standing of the company, and evaluate the C-level executives. But in the end, there are no guarantees. It is often your gut that will push you to invest in Able, Inc., instead of Baker Corp. For a *seasoned* investor, listening to the gut to tip the scales toward a particular investment will pay off more often because of an accumulated lifetime of gathering data about the intangibles that make a good investment. The key to whether a gut feeling is likely to be a good one or not is *experience*.

Gut feelings are analogous to the sound processing strategies our sound mind has honed to a default state, ready to react to sound or play a long-neglected concerto based on our accumulated experience. Much of our perception of the world is as intangible as a gut feeling. Yet by analyzing how sound is processed by the signals inside the head, we can glimpse how our experience has shaped our interpretation of the sounds outside the head. Everyone has their own sonic finger-print. How have the faders on our mixing board been adjusted? Have we honed our sound minds by making music or learning a second language? Have we dulled our brain with noise exposure or deprived it of the rich sounds of language? Which choices can we make now to allow sound to change us for the better as we live out our lives?

The Sound Mind Shapes Our Choices for Our Sonic Future

The power of sound is a well-kept secret. My goal in writing this book has been to *give voice* to the power of sound. A call to shift, if only occasionally, away from our visually dominated and material-istic outlook toward what sound can offer. Armed with this knowl-edge, we can recognize sound as an ally in our own lives, the lives of others, and other living things.

Who we are and what we value impact the world we live in. The sound minds we develop sculpt our sonic world based on what we prize and what we dislike. And our choices will affect the sonic world of our children and our children's children. Our sound mind decisions can lead us to priorities for how we lead our lives.

Before I close this book, I would like to share some of the sound mind choices I have made for myself and my family, and leave you with some ideas to ponder about leading life with a sound mind:

- When they were little, I enforced three rules for my sons. They had to take their schoolwork seriously. They had to tell me where

Figure 14.2
The sound mind guides our choices for our future sonic world.

they were, always. And they had to practice their instruments. If they did those three things, they had every freedom. Today, although not one of them is a professional musician, they speak the language of music well enough to make music on their own and with others. I prize the times we make music together.

- Large parts of the world are adopting English as a de facto lingua franca. Our sound minds evolve from the languages we speak. Might this help us understand each other better? How does being familiar with the sounds of more than one language predispose us to feel about each other?

- When we have trouble reading road signs, we know it's time to get our eyes checked. Losing hearing is much more subtle because it's all too easy to blame the sound. It would never cross your mind to accuse the department of transportation of erecting signs with blurry letters. But you might accuse someone of mumbling. As we get older and "people start mumbling," hearing aids can help keep the input to the sound mind sharp.

- What does our lo-fi music experience do to the sound mind? We have moved toward ever-more-compressed file delivery (streaming, MP3s) and listening through smartphone speakers. A music teacher told me many of his students could not distinguish between music played through a hi-fi system and the same music played through a smartphone. If the brain has adapted to a tinny simulacrum of music, will it ever learn to hear the rich sounds that exist? Are we becoming less inclined to seek live music, listen with high-quality speakers, and create architectural spaces that preserve sound ingredients? Will these ingredients become lost to us? Will we make less interesting music? Some musicians including Linda Ronstadt, Brian Eno, Kate Bush, and late-period Beatles chose to perform in smaller venues or in the studio in order to deliver their music as they valued it most, ungarbled by overamplified sound ricocheting off stadium rafters. They listened to their sound minds.

- What is the environmental impact of indifference to noise? A person who values the sounds of the wilderness will want to preserve the privilege of hearing that soundscape. Someone who is less aware of sound might see wilderness as an economic opportunity. "Scenic helicopter tours, seventy-five dollars for thirty minutes."

- Some students tell me they choose to study in a coffee shop or with the TV on in the background. They claim having to ignore the sound helps them concentrate. I often learn they grew up in a noisy place. From a young age, their hearing brains were trained to crave noise to be productive. What happens to the sound mind if our default brain networks regard sound as something to ignore?

- The purpose of music is connection but it has become a pervasive background feature in most public settings. What if music becomes something you increasingly tune out?

- Noise begets stress and stress begets noise. Maybe you are feeling some stress and you react by stomping around the house.

This ups the noise level and your roommate turns up the volume on the TV to compensate. The loud TV annoys you further and you stomp louder. This type of noise-induced positive feedback loop has been studied and, sure enough, people exposed to noise become more aggressive and eager to zap fellow research participants with electric shocks.[13] How do you feel about the person who just honked their horn at you?

- The sound mind is among the victims of a concussion. Athletes as much as anyone rely on making sense of sound to do their job. Getting back on the field might seem less urgent if an athlete realizes he will not be at his best if his sound mind is underperforming.

- What can we do, as advocates of the sound mind, to influence urban planning? How do we ensure that our built environment provides the best opportunity to listen clearly so we can think, learn, and communicate with each other? When we consider the ecological impact of the world we create—housing, commerce, transportation, or otherwise—we often focus on sustainability, environmental sensitivity, and aesthetic visual appeal. Are we willing to pay more for quieter air-conditioning, heating systems, subways . . . for better sound aesthetics?

- Texts and emails are rapidly supplanting phone calls. With this change, information can be conveyed, but context suffers. We have all experienced someone misinterpreting sarcasm for anger or a casual request for an urgent one. Peppering text messages with emoji only goes so far. Are we gradually failing to develop a sensitivity to "tone of voice" as voice communication diminishes?

- When we *do* make a phone call to a business, we usually have to find our way to the right department through a series of computerized voice menus. Are our sound minds being dulled in the ability to pick up on voice-conveyed nuance by our increasing exposure to nuance-free speech?

- Perhaps the sound mind can teach us about big questions biologists and philosophers have been grappling with for centuries. What is consciousness? What is the nature of "self"? What connection do we have with the world? What is the nature of spirituality, the nature of memory, the intersection of brain, body, and mind?

Biologically speaking, we are what we do. We are what we pay attention to and how we spend our time. We are what moves us. We are what we love.

What I have shared with you in this book are my scientific gut feelings based on years of thinking about the biology of hearing. Science cannot furnish every answer, but we have abundant evidence to trust that sound is a force shaping our minds. We can give voice to the power of sound by considering initiatives for making music, foreign-language learning, and athletics. Sound has a place in medicine for people (and coral reefs). We can work to honor silence, the sounds of home, the soft sounds we love, and avoid excessive noise in the places we spend our time. We can consider sound in the creation of new spaces. We can try to make music with our families and friends. We can appreciate the beauty of sound with awe.

Acknowledgments

This book would not exist without Trent Nicol. He has been my partner in every stage of writing this book, as he has been my lab partner at Brainvolts for the past thirty years. Trent is able to give voice to my ideas when they are not yet words in my mind. Trent articulates my intentions more gracefully than I can. Often I read his words and think, "Yes, that's what I meant to say." Plus, Trent manages to be clever and funny in ways I wish I could be. Just as Trent's contributions at Brainvolts span the majestic to the mundane, he also compiled the references for this book and constructed the graphs and simple figures. Finally, as a restorer of old radios, Trent is responsible for the uncommonly sweet sounds I listen to each day coming from the radios, his radios, that live in my office and in our kitchen.

When I began working on this book, I didn't know I needed an agent. I didn't even know what a book agent was. Agent Anne, as I affectionately call Anne Edelstein, is the rock upon which my book writing experience is grounded. Anne has led me, patiently and responsively, teaching me every step of the way. Her first email— "Just to let you know I'm riveted by your proposal and chapters, which I've been reading aloud in the car to my husband on our way up to Maine"—was just what I needed to feel like I was on the right track. I had not encountered "content editing" before, but was fortunate to experience it firsthand through Anne's thoughtful perspective, her reorganization of the narrative, and her fine use of words. As countless book-related issues crop up I continue to be grateful and relieved when Anne tells me, "I'll take care of that."

Katie Shelly had the job of (literally) bringing out the art in science. Katie drew and conceived most of the illustrations in this book. Beyond the beauty and imagination palpable in her art, Katie's responsivity, from another continent (she's in Spain), often exceeds that of colleagues working down the hall. Her endless flexibility, patience, and receptivity to my many suggestions and criticisms have made Katie fun to work with. On many occasions Katie has managed to focus my disorganized requests into imaginative, creative, and actionable ideas that manifest as designs way beyond my expectations.

Hannah Geil-Neufeld read every word of my first draft of the book. As a representative of my target audience, a thoughtful and curious reader, she pointed out what was hard to understand, and what was too technical or assumed too much scientific background. Her requests for explanation guided my thinking and my writing. I am grateful for Hannah's editing; she is a beautiful writer and fun to work with.

Many thanks go to my editor Robert Prior, for his confidence in welcoming the book to the MIT Press. His insightful suggestions of my chapter headings helped establish the tone of the book from the start. I am grateful to production editor Judith Feldmann, art coordinator Sean Reilly, assistant acquisitions editor Anne-Marie Bono, and publicist Angela Baggetta. I thank the anonymous reviewers solicited by the MIT Press—let me know who you are so I can thank you personally.

Thank you to everyone who commented on early versions of the book: Dan Rocker, Jennifer Krizman, Travis White-Schwoch, Silvia Bonacina, Rembrandt Otto-Meyer, Graham Straus, Curt and Linda Matthews, and Salvatore Spina.

While writing this book I needed an expert on the biology of learning to tell me when I was barking up the wrong tree, and thank goodness I was able to turn to Kasia Bieszczad. Kasia is a neuroscientist who rigorously approaches auditory learning at a cellular level and she is an educator who values the teaching of complex ideas so they can be understood by anyone interested in knowing them. Her feedback was invaluable.

My hot dog stand wouldn't exist if not for my many scientist collaborators and mentors, past and present. Some outfitted me with equipment to fire up the grill, others provided me with an ever-expanding pantry full of buns, condiments, and fries that keep the customers coming.

Many years ago, Raymond Carhart was kind enough to listen to a clueless student, take me by the hand, and introduce me to Peter Dallos down the hall. Later mentors include John Disterhoft, Laszlo Stein, Earleen Elkins, and Ed Rubel.

I would like to thank Brainvolts alumni doctoral students Anu Sharma, Cynthia King, Kelly Tremblay, Jenna Cunningham, Brad Wible, Jill Firszt, Erin Hayes, Gabriella Musacchia, Krista Johnson, Dan Abrams, Nicole Russo, Jade Wang, Judy Song, Kyung Myun Lee, Jane Hornickel, Samira Anderson, Erika Skoe, Dana Strait, Karen Chan, Alexandra Parbery-Clark, Jennifer Krizman, Jessica Slater, and Elaine C. Thompson. I'd like to thank postdoctoral fellows Alan Micco, Thomas Littman, Anu Sharma, Elizabeth Dinces, Ann Bradlow, Ivy Dunn, Catherine Warrier, Lauri Olivier, Karen Banai, Frederic Marmel, Bharath Chandrasekaran, Yun Nan, Jason Thompson, Erika Skoe, Dana Strait, Adam Tierney, Ahren Fitzroy, and Spencer Smith. And I'd like to thank the dozens of undergraduates, high school students, and clinical doctorate students as well as Bob Conway for his creative attention to our work space. My colleague Therese McGee launched our first discussions on using subcortical neural synchrony to get at speech sound processing in humans. Without her, who knows what I'd have been doing the last twenty-five years. And speaking of people who built the foundation, thanks to Jim Perkins for a beautiful family.

In my current research, I am grateful for Jennifer Krizman, the leader of several ongoing projects, polymath Travis White-Schwoch, rhythm specialist Silvia Bonacina, whose Italian accent reminds me of la mamma—sound's connection to home, and Rembrandt Otto-Meyer who keeps data collection going despite the pandemic. Jen created our collective labor of love, the Brainvolts website, updated almost daily by Rembrandt. Brainvolts colleagues are colleagues for life due to the nature of science, the longterm connections we build, and the natural opportunities to see each other at conferences. Jen, Erika, and Travis's many strengths include their ability to throw Brainvolts gatherings, as they did famously at a recent international conference. Home away from home.

Brainvolts' science crucially depends on our collaborators in education, music, biology, athletics, medicine, and industry—people who operate in the world outside the lab, the world I want our science to live in. Special thanks to Margaret Martin, Kate Johnston, Tory Lindley, Cynthia LaBella, Daniele Colegrove, Jeff Mjaanes, Ann Bradlow, Tom Carrell, and Steve Zecker. Backing one another up with our various modes of operation makes what we do effective and fun.

Renée Fleming, Mickey Hart, and Zakir Hussain are my models for how to make art work for the good of science. Thank you for allowing me to do a little of this with you. Thank you Arne Starr for sharing your art in figures 2.3, 2.4, and 2.6.

Brainvolts has enjoyed continuous federal funding through the National Science Foundation and many institutes of the National Institutes of Health—Child Health and Human Development, Mental Health, Neurological Disorders and Stroke, Deafness and Communication Disorders, and Aging. We are grateful for support from foundations including The American Hearing Research Foundation, The Cade Royalty Fund, The Dana Foundation, The G. Harold & Leila Y. Mathers Foundation, The Hunter Family Foundation,

The Rachel E. Golden Foundation, The Spencer Foundation, The National Academy of Recording Arts and Sciences, The National Association of Music Merchants, and The National Operating Committee on Standards for Athletic Equipment. We have been lucky to have commercial support from Med-El, Interactive Metronome, and Phonak. Thanks to the Knowles Hearing Center and Northwestern University for giving Brainvolts a home.

Biggest thanks go to my family. My parents nurtured my sound mind from the beginning. Nick Friedman, Leah Campbell, Hannah Geil-Neufeld, Grant Dawson, Susie Richard, Lucio Sadoch and Lynn McNutt have maintained their enthusiasm for my project despite my inability to stop talking about it. They have provided substantive comments throughout, cheering me to the finish. Thank you to Brainvolts' guardian angel Bic Wirtz, who is my best friend. Everyone needs a best friend and I got the best one. Mikey, Russell, and Nick Perkins, and Marshall Dawson to whom this book is dedicated deserve special mention.

My son Nick is a daily reminder of the sensory-mind partnership. Nick the chef creates nourishing and scrumptious meals. With his knowledge of food, flavor, ingredients, and kitchen chemistry, he makes everything taste good, even dishes with milk and butter despite his life-threatening allergy to dairy products. We call him "our Beethoven."

My son Mikey helps me to remember to prize the concepts of home, belonging, and community. He honors these ideas by building places to live in that encourage connection by using wood—a living substance. He challenges me to think about how sound fits into these contexts. In living life according to one's values, Mikey is the most uncompromising person I know.

Nick and Mikey have created work places built on a sense of belonging. These endeavors are a constant source of inspiration for Brainvolts, as science at its best is collaborative, integrative, and cumulative.

At once meticulous and able to entertain infinite possibilities, my son Russell personifies a common thread between art and science. As an artist, scholar, and musician, he gives the gift of his thoughts in his kindness to others. Russell has disciplined himself to learn since childhood. His love of learning for its own sake, and because it feels right, is the best foundation for art and science I can think of.

Nothing is better for my sound mind than living with my husband Marshall. He shows me what the sound mind is capable of, from recognizing cartoon characters in the voices of actors, to how he lives his life as a note reading, imitating, and improvising musician. He graces the sonic world with his teaching and performance. In writing this book and always, really, I am grateful I can count on Marshall to be unmercifully honest. And, I am glad he finds microphones and electrical impedance suitable topics for dinner conversation.

Glossary

Afferent A movement toward a node (e.g., the brain). In the auditory system, it is the progression from cochlea, through *midbrain* and thalamus to auditory cortex.

Amplitude modulation (AM) A fluctuation of sound intensity (loud-soft-loud-soft, like many alarms). The vibration of our vocal cords as they open and close amplitude modulates the sound. The AM rate is the *fundamental frequency* of a speech sound, which contributes to the *pitch* of the voice (lower in men, higher in women).

Binding problem How input from multiple sensory systems are combined and coordinated to form a unified object.

Efferent A movement away from a node (e.g., the brain). In the auditory system, it involves, for example, neural signals from cortex to thalamus, from *midbrain* to cochlea, etc.

Frequency A count of something per a fixed time unit. The frequency of a sound, measured in cycles per second (hertz), determines its *pitch*.

Frequency following response (FFR) A *neurophysiological* response to sound that reflects how the brain processes the many ingredients in sound such as *pitch*, timing, and *timbre*.

FM sweep A frequency modulated (FM) sound changes in *frequency* over time. An example is sweeping across the notes on a piano or a siren. The FM sweep is an important ingredient of speech, especially pertinent in consonant sounds in the form of bands of concentrated acoustic energy sweeping from low to high frequencies or vice versa.

Fundamental frequency The lowest frequency of a harmonic sound. The fundamental frequency confers the perception of *pitch*.

Hair cell Located in the inner ear, hair cells are gently moved by fluids set in motion by air movement (i.e., sound). This movement triggers an electrical signal, thereby completing the *transduction* from sound to electricity.

Harmonics *Frequencies* present in a sound at integer multiples above the *fundamental frequency*. For example, a sound with a fundamental frequency of 150 Hz will have harmonics at 300, 450, 600, etc., Hz.

Hyperacusis A condition where low- or moderate-intensity sounds are perceived to be uncomfortably loud.

Inhibition A process of suppressing the firing of neurons below their spontaneous firing level. As an example, in the auditory system, neurons tuned to frequencies adjacent to an incoming sound will reduce their firing so that the firing of exactly tuned neurons is emphasized.

Limbic system The brain network that support emotion, motivation, and feelings such as pleasure.

Midbrain A region of the brain that lies between the brainstem and the cortex. In the auditory system, the midbrain, a hub at the crossroads of the sensory, motor, cognitive, and reward systems, is a particularly useful window into the sound mind.

Mismatch negativity A *neurophysiological* response to a change in an ongoing sound pattern. For example, the movement of a snake in the grass creates a change in the ongoing sound of rustling grass.

Misophonia A condition whereby sounds such as chewing or a ticking clock are excessively disturbing.

Neural plasticity The ability of neurons in the brain to change their responsivity due to learning. A classic example is the expansion of the somatosensory and motor maps corresponding to the left fingers of a violinist.

Neural synchrony When neurons fire together to the timing landmarks in the sound.

Neuroeducation Also known as educational neuroscience. A science-based approach to informing teaching methods that maximize academic achievement in children.

Neurophysiology The study of the function of the nervous system.

Otoacoustic emission Sounds emitted from the ear that can be used to assess function of the outer *hair cells* and the *efferent* control of them.

Phaselocking A repetitive firing of a neuron to a repeating, cyclical auditory signal such as a sinewave or a jackhammer.

Phoneme The smallest sound unit of speech. A phoneme is not mapped one-to-one with letters. For example, the phoneme /f/, one of forty-four in English, can be found in the words fact, phone, half, and laugh.

Pitch The perceptual sensation of sound *frequency*. Generally, a sound with a high frequency will sound high in pitch. A sound with low frequency will sound low in pitch.

Reticular activating system Brain centers involved in arousal and attention.

Spectral shape The pattern of *harmonic* energy in a sound that gives rise to the perception of *timbre*. The spectral shape of a speech sound determines which consonant or vowel was spoken. In music, it determines which instrument was played.

Spectrum A visualization of the *frequencies* that make up a sound or brain signal. A spectrogram is a visualization of frequencies as they change over time.

Timbre The quality of a sound imparted by its *spectral shape*. An oboe and a trombone have different timbres even when playing the same note. Although we rarely talk about the timbre of speech, the same principle that distinguishes among musical instruments applies to distinguishing speech sounds such as "ah" and "oo."

Tonotopy The tendency for structures in the auditory pathway to be arranged topographically by preferred frequency. Also called tonotopicity by presumably the same people who use words like beautifulness instead of beauty.

Transduce To convert something to another form. As used in this book, the air pressure waves of sound are converted by the cochlea to electricity.

Working memory A temporary form of memory that can be accessed and manipulated. To contrast echoic memory from working memory, it is the difference between rote repetition of five heard words and repeating them in alphabetical order.

Notes

Introduction

1. E. H. Lenneberg, *Biological Foundations of Language* (New York: Wiley, 1967).

2. D. Harris, P. Dallos, and N. Kraus, "Forward and Simultaneous Tonal Suppression of Single-Fiber Responses in the Chinchilla Auditory Nerve," *Journal of the Acoustical Society of America* 60 (1976): S81.

3. N. Kraus and J. F. Disterhoft, "Response Plasticity of Single Neurons in Rabbit Auditory Association Cortex during Tone-Signalled Learning," *Brain Research* 246, no. 2 (1982): 205–215.

4. A. W. Scott, N. M. Bressler, S. Ffolkes, J. S. Wittenborn and J. Jorkasky, "Public Attitudes about Eye and Vision Health," *JAMA Ophthalmology* 134, no. 10 (2016): 1111–1118.

5. F. R. Lin and M. Albert, "Hearing Loss and Dementia—Who Is Listening?" *Aging & Mental Health* 18, no. 6 (2014): 671–673.

6. A. Krishnan, Y. S. Xu, J. Gandour, and P. Cariani, "Encoding of Pitch in the Human Brainstem Is Sensitive to Language Experience," *Cognitive Brain Research* 25, no. 1 (2005): 161–168.

7. N. Kraus & T. Nicol "The Power of Sound for Brain Health," *Nature Human Behaviour* 1 (2017): 700-702.

Chapter 1

1. T. D. Hanley, J. C. Snidecor, and R. L. Ringel, "Some Acoustic Differences among Languages," *Phonetica* 14 (1966): 97–107; A. B. Andrianopoulos, K. N. Darrow, and

J. Chen, "Multimodal Standardization of Voice among Four Multicultural Populations: Fundamental Frequency and Spectral Characteristics," *Journal of Voice* 15, no. 2 (2001): 194–219.

2. S. A. Xue, R. Neeley, F. Hagstrom, and J. Hao, "Speaking F0 Characteristics of Elderly Euro-American and African-American Speakers: Building a Clinical Comparative Platform," *Clinical Linguistics & Phonetics* 15, no. 3 (2001): 245–252.

3. B. Lee and D. V. L. Sidtis, "The Bilingual Voice: Vocal Characteristics when Speaking Two Languages across Speech Tasks," *Speech, Language and Hearing* 20, no. 3 (2017): 174–185.

Chapter 2

1. R. Wallace, *Hearing Beethoven: A Story of Musical Loss and Discovery* (Chicago: The University of Chicago Press, 2018).

2. J. Cunningham, T. Nicol, C. D. King, S. G. Zecker, and N. Kraus, "Effects of Noise and Cue Enhancement on Neural Responses to Speech in Auditory Midbrain, Thalamus and Cortex," *Hearing Research* 169 (2002): 97–111.

3. E. M. Ostapoff, J. J. Feng, and D. K. Morest, "A Physiological and Structural Study of Neuron Types in the Cochlear Nucleus. II. Neuron Types and Their Structural Correlation with Response Properties," *Journal of Comparative Neurology* 346, no. 1 (1994): 19–42.

4. J. J. Feng, S. Kuwada, E. M. Ostapoff, R. Batra, and D. K. Morest, "A Physiological and Structural Study of Neuron Types in the Cochlear Nucleus. I. Intracellular Responses to Acoustic Stimulation and Current Injection," *Journal of Comparative Neurology* 346, no. 1 (1994): 1–18.

5. Source for figure 2.5: N. B. Cant, "The Cochlear Nucleus: Neuronal Types and Their Synaptic Organization," in *The Mammalian Auditory Pathway: Neuroanatomy*, ed. D. B. Webster, A. N. Popper, and R. R. Fay, Springer Handbook of Auditory Research (Springer-Verlag, 1992), 66–119.

6. R. D. Frisina, R. L. Smith, and S. C. Chamberlain, "Encoding of Amplitude Modulation in the Gerbil Cochlear Nucleus: I. A Hierarchy of Enhancement," *Hearing Research* 44, no. 2–3 (1990): 99–122.

7. T. C. T. Yin, "Neural Mechanisms of Encoding Binaural Localization Cues in the Auditory Brainstem," in *Integrative Functions in the Mammalian Auditory Pathway*, ed. D. Oertel, R. R. Fay, and A. N. Popper, Springer Handbook of Auditory Research (New York: Springer, 2002).

8. C. E. Schreiner and G. Langner, "Periodicity Coding in the Inferior Colliculus of the Cat. II. Topographical Organization," *Journal of Neurophysiology* 60, no. 6 (1988):

1823–1840; G. Langner, M. Albert, and T. Briede, "Temporal and Spatial Coding of Periodicity Information in the Inferior Colliculus of Awake Chinchilla (*Chinchilla laniger*)," *Hearing Research* 168, no. 1–2 (2002): 110–130.

9. G. M. Shepherd, *Neurogastronomy: How the Brain Creates Flavor and Why It Matters* (New York: Columbia University Press, 2012).

10. G. H. Recanzone, D. C. Guard, M. L. Phan, and T. K. Su, "Correlation between the Activity of Single Auditory Cortical Neurons and Sound-Localization Behavior in the Macaque Monkey," *Journal of Neurophysiology* 83, no. 5 (2000): 2723–2739; J. C. Middlebrooks and J. D. Pettigrew, "Functional Classes of Neurons in Primary Auditory Cortex of the Cat Distinguished by Sensitivity to Sound Location," *Journal of Neuroscience* 1, no. 1 (1981): 107–120.

11. L. Feng and X. Wang, "Harmonic Template Neurons in Primate Auditory Cortex Underlying Complex Sound Processing," *Proceedings of the National Academy of Sciences of the United States of America* 114, no. 5 (2017): E840–848.

12. Y. I. Fishman, I. O. Volkov, M. D. Noh, P. C. Garell, H. Bakken, J. C. Arezzo, M. A. Howard, and M. Steinschneider, "Consonance and Dissonance of Musical Chords: Neural Correlates in Auditory Cortex of Monkeys and Humans," *Journal of Neurophysiology* 86, no. 6 (2001): 2761–2788; M. J. Tramo, J. J. Bharucha, and E. E. Musiek, "Music Perception and Cognition Following Bilateral Lesions of Auditory Cortex," *Journal of Cognitive Neuroscience* 2, no. 3 (1990): 195–212; I. Peretz, A. J. Blood, V. Penhune, and R. Zatorre, "Cortical Deafness to Dissonance," *Brain* 124, no. 5 (2001): 928–940.

13. A. Bieser and P. Muller-Preuss, "Auditory Responsive Cortex in the Squirrel Monkey: Neural Responses to Amplitude-Modulated Sounds," *Experimental Brain Research* 108, no. 2 (1996): 273–284; H. Schulze and G. Langner, "Periodicity Coding in the Primary Auditory Cortex of the Mongolian Gerbil (Meriones Unguiculatus): Two Different Coding Strategies for Pitch and Rhythm?" *Journal of Comparative Physiology A: Neuroethology, Sensory, Neural, and Behavioral Physiology* 181, no. 6 (1997): 651–663.

14. C. T. Engineer, C. A. Perez, Y. H. Chen, R. S. Carraway, A. C. Reed, J. A. Shetake, V. Jakkamsetti, K. Q. Chang, and M. P. Kilgard, "Cortical Activity Patterns Predict Speech Discrimination Ability," *Nature Neuroscience* 11, no. 5 (2008): 603–608.

15. P. Heil and D. R. Irvine, "First-Spike Timing of Auditory-Nerve Fibers and Comparison with Auditory Cortex," *Journal of Neurophysiology* 78, no. 5 (1997): 2438–2454.

16. R. C. deCharms, D. T. Blake, and M. M. Merzenich, "Optimizing Sound Features for Cortical Neurons," *Science* 280, no. 5368 (1998): 1439–1443.

17. A. S. Bregman, *Auditory Scene Analysis: The Perceptual Organization of Sound* (Cambridge, MA: MIT Press, 1990).

18. L. J. Hood, C. I. Berlin, and P. Allen, "Cortical Deafness: A Longitudinal Study," *Journal of the American Academy of Audiology* 5, no. 5 (1994): 330–342.

19. G. Vallortigara, L. J. Rogers, and A. Bisazza, "Possible Evolutionary Origins of Cognitive Brain Lateralization," *Brain Research Reviews* 30, no. 2 (1999): 164–175.

20. R. J. Zatorre, A. C. Evans, E. Meyer, and A. Gjedde, "Lateralization of Phonetic and Pitch Discrimination in Speech Processing," *Science* 256, no. 5058 (1992): 846–849; M. J. Tramo, G. D. Shah, and L. D. Braida, "Functional Role of Auditory Cortex in Frequency Processing and Pitch Perception," *Journal of Neurophysiology* 87, no. 1 (2002): 122–139.

21. I. McGilchrist, *The Master and His Emissary: The Divided Brain and the Making of the Western World* (New Haven: Yale University Press, 2009).

22. N. Kraus and T. Nicol, "Brainstem Origins for Cortical 'What' and 'Where' Pathways in the Auditory System," *Trends in Neurosciences* 28 (2005): 176–181.

23. A. Starr, T. W. Picton, W. Sininger, L. J. Hood, and C. I. Berlin, "Auditory Neuropathy," *Brain* 119, no. 3 (1996): 741–753; N. Kraus, Ö. Özdamar, L. Stein, and N. Reed, "Absent Auditory Brain Stem Response: Peripheral Hearing Loss or Brain Stem Dysfunction?" *Laryngoscope* 94: (1984): 400–406.

24. M. N. Wallace, R. G. Rutkowski, and A. R. Palmer, "Identification and Localisation of Auditory Areas in Guinea Pig Cortex," *Experimental Brain Research* 132, no. 4 (2000): 445–456.

25. N. Kraus and T. White-Schwoch, "Unraveling the Biology of Auditory Learning: A Cognitive-Sensorimotor-Reward Framework," *Trends in Cognitive Sciences* 19 (2015): 642–654; N. M. Weinberger, "The Medial Geniculate, Not the Amygdala, as the Root of Auditory Fear Conditioning," *Hearing Research* 274, no. 1–2 (2001): 61–74; E. Hennevin, C. Maho, and B. Hars, "Neuronal Plasticity Induced by Fear Conditioning Is Expressed During Paradoxical Sleep: Evidence from Simultaneous Recordings in the Lateral Amygdala and the Medial Geniculate in Rats," *Behavorial Neuroscience* 112, no. 4 (2008): 839–862.

26. E. D. Jarvis, "Learned Birdsong and the Neurobiology of Human Language," *Annals of the New York Academy of Sciences* 1016 (2004): 749–777.

27. M. H. Giard, L. Collet, P. Bouchet, and J. Pernier, "Auditory Selective Attention in the Human Cochlea," *Brain Research* 633, no. 1–2 (1994): 353–356.

28. M. Ahissar and S. Hochstein, "The Reverse Hierarchy Theory of Visual Perceptual Learning," *Trends in Cognitive Sciences* 8, no. 10 (2004): 457–464.

29. M. Schutz and S. Lipscomb, "Hearing Gestures, Seeing Music: Vision Influences Perceived Tone Duration," *Perception* 36, no. 6 (2007): 888–897.

30. R. Gillespie, "Rating of Violin and Viola Vibrato Performance in Audio-Only and Audiovisual Presentations," *Journal of Research in Music Education* 45, no. 2 (1997): 212–220.

31. H. Saldaña and L. D. Rosenblum, "Visual Influences on Auditory Pluck and Bow Judgments," *Perception and Psychophysics* 54, no. 3 (1993): 406–416.

32. H. McGurk and J. MacDonald, "Hearing Lips and Seeing Voices," *Nature* 264, no. 5588 (1976): 746–748.

33. J. A. Grahn and M. Brett, "Rhythm and Beat Perception in Motor Areas of the Brain," *Journal of Cognitive Neuroscience* 19, no. 5 (2007): 893–906.

34. A. Lahav, E. Saltzman, and G. Schlaug, "Action Representation of Sound: Audio-motor Recognition Network While Listening to Newly Acquired Actions," *Journal of Neuroscience* 27, no. 2 (2007): 308–314; J. Haueisen and T. R. Knosche, "Involuntary Motor Activity in Pianists Evoked by Music Perception," *Journal of Cognitive Neuroscience* 13, no. 6 (2001): 786–792.

35. B. Haslinger, P. Erhard, E. Altenmuller, U. Schroeder, H. Boecker, and A. O. Ceballos-Baumann, "Transmodal Sensorimotor Networks during Action Observation in Professional Pianists," *Journal of Cognitive Neuroscience* 17, no. 2 (2005): 282–293; G. A. Calvert, E. T. Bullmore, M. J. Brammer, R. Campbell, S. C. Williams, P. K. McGuire, P. W. Woodruff, S. D. Iversen, and A. S. David, "Activation of Auditory Cortex During Silent Lipreading," *Science* 276, no. 5312) (1997): 593–696.

36. B. W. Vines, C. L. Krumhansl, M. M. Wanderley, D. J. Levitin, "Cross-modal Interactions in the Perception of Musical Performance," *Cognition* 101, no. 1 (2006): 80–103; C. Chapados, D. J. Levitin, "Cross-modal Interactions in the Experience of Musical Performances: Physiological Correlates," *Cognition* 108, no. 3 (2008): 639–651; B. W. Vines, C. L. Krumhansl, M. M. Wanderley, I. M. Dalca, and D. J. Levitin, "Music to My Eyes: Cross-modal Interactions in the Perception of Emotions in Musical Performance," *Cognition* 118, no. 2 (2011): 157–170.

37. E. Kohler, C. Keysers, M. A. Umilta, L. Fogassi, V. Gallese, and G. Rizzolatti, "Hearing Sounds, Understanding Actions: Action Representation in Mirror Neurons," *Science* 297, no. 5582 (2002): 846–848; V. Gallese, L. Fadiga, L. Fogassi, and G. Rizzolatti, "Action Recognition in the Premotor Cortex," *Brain* 119, no. 2 (1996): 593–609.

38. L. M. Oberman, E. M. Hubbard, J. P. McCleery, E. L. Altschuler, V. S. Ramachandran, and J. A. Pineda, "EEG Evidence for Mirror Neuron Dysfunction in Autism Spectrum Disorders," *Brain Research: Cognitive Brain Research* 24, no. 2 (2005): 190–198; G. Hickok, *The Myth of Mirror Neurons: The Real Neuroscience of Communication and Cognition* (New York: W. W. Norton, 2014).

39. S. Montgomery, *The Soul of an Octopus: A Surprising Exploration into the Wonder of Consciousness* (New York: Atria Books, 2015).

40. J. Panksepp, *Affective Neuroscience: The Foundations of Human and Animal Emotions* (New York: Oxford University Press, 1998).

41. L. Selinger, K. Zarnowiec, M. Via, I. C. Clemente, and C. Escera, "Involvement of the Serotonin Transporter Gene in Accurate Subcortical Speech Encoding," *Journal of Neuroscience* 36, no. 42 (2016): 10782–10790; L. M. Hurley and G. D. Pollak, "Serotonin Differentially Modulates Responses to Tones and Frequency-Modulated Sweeps in the Inferior Colliculus," *Journal of Neuroscience* 19, no. 18 (1999): 8071–8082; L. M. Hurley and G. D. Pollak, "Serotonin Effects on Frequency Tuning of Inferior Colliculus Neurons," *Journal of Neurophysiology* 85, no. 2 (2001): 828–842; J. A. Schmitt, M. Wingen, J. G. Ramaekers, E A. Evers, and W. J. Riedel, "Serotonin and Human Cognitive Performance," *Current Pharmaceutical Design* 12, no. 20 (2006): 2473–2486; A. G. Fischer and M. Ullsperger, "An Update on the Role of Serotonin and Its Interplay with Dopamine for Reward," *Frontiers in Human Neuroscience* 11 (2017): 484.

42. B. J. Marlin, M. Mitre, J. A. D'Amour, M. V. Chao, and R. C. Froemke, "Oxytocin Enables Maternal Behaviour by Balancing Cortical Inhibition," *Nature* (2015), https:doi.org/10.1038/nature14402.

Chapter 3

1. W. Penfield and E. Boldrey, "Somatic Motor and Sensory Representation in the Cerebral Cortex of Man as Studied by Electrical Stimulation," *Brain* 60 (1937): 389–443; J. L. Hampson, C. R. Harrison, and C. N. Woolsey, "Somatotopic Localization in the Cerebellum," *Federation Proceedings* 5, no. 1 (1946): 41.

2. M. M. Merzenich, J. H. Kaas, J. Wall, R. J. Nelson, M. Sur, and D. Felleman, "Topographic Reorganization of Somatosensory Cortical Areas 3b and 1 in Adult Monkeys Following Restricted Deafferentation," *Neuroscience* 8, no. 1 (1983): 33–55.

3. M. M. Merzenich, P. L. Knight, and G. L. Roth, "Representation of Cochlea Within Primary Auditory Cortex in the Cat," *Journal of Neurophysiology* 38, no. 2 (1975): 231–249.

4. C. A. Atencio, D. T. Blake, F. Strata, S. W. Cheung, M. M. Merzenich, and C. E. Schreiner, "Frequency-Modulation Encoding in the Primary Auditory Cortex of the Awake Owl Monkey," *Journal of Neurophysiology* 98, no. 4 (2007): 2182–2195; G. H. Recanzone, C. E. Schreiner, M. L. Sutter, R. E. Beitel, and M. M. Merzenich, "Functional Organization of Spectral Receptive Fields in the Primary Auditory Cortex of the Owl Monkey," *Journal of Comparative Neurology* 415, no. 4 (1999): 460–481.

5. G. H. Recanzone, C. E. Schreiner, and M. M. Merzenich, "Plasticity in the Frequency Representation of Primary Auditory Cortex Following Discrimination Training in Adult Owl Monkeys," *Journal of Neuroscience* 13, no. 1 (1993): 87–103; M. M. Merzenich, P. L. Knight, and G. L. Roth, "Representation of Cochlea Within Primary Auditory Cortex in the Cat," *Journal of Neurophysiology* 38, no. 2 (1975): 231–234; J. S. Bakin and N. M. Weinberger, "Classical Conditioning Induces Cs-Specific Receptive-Field Plasticity in the Auditory Cortex of the Guinea Pig," *Brain Research*

536, no. 1–2 (1990): 271–286; K. M. Bieszczad, A. A. Miasnikov, and N. M. Weinberger, "Remodeling Sensory Cortical Maps Implants Specific Behavioral Memory," *Neuroscience* 246 (2013): 40–51; M. Brown, D. R. Irvine, and V. N. Park, "Perceptual Learning on an Auditory Frequency Discrimination Task by Cats: Association with Changes in Primary Auditory Cortex," *Cerebral Cortex* 14, no. 9 (2004): 952–965; J. M. Edeline, and N. M. Weinberger. "Receptive Field Plasticity in the Auditory Cortex During Frequency Discrimination Training: Selective Retuning Independent of Task Difficulty," *Behavioral Neuroscience* 107, no. 1 (1993): 82–103; G. A. Elias, K. M. Bieszczad, and N. M. Weinberger, "Learning Strategy Refinement Reverses Early Sensory Cortical Map Expansion but Not Behavior: Support for a Theory of Directed Cortical Substrates of Learning and Memory," *Neurobiology of Learning and Memory* 126 (2015): 39–55.

6. B. Röder, O. Stock, S. Bien, H. Neville, and F. Rösler, "Speech Processing Activates Visual Cortex in Congenitally Blind Humans," *European Journal of Neuroscience* 16, no. 5 (2002): 930–936.

7. N. Sadato, A. Pascual-Leone, J. Grafman, V. Ibanez, M. P. Deiber, G. Dold, and M. Hallett, "Activation of the Primary Visual Cortex by Braille Reading in Blind Subjects," *Nature* 380, no. 6574 (1996): 526–528.

8. H. Nishimura, K. Hashikawa, K. Doi, T. Iwaki, Y. Watanabe, H. Kusuoka, T. Nishimura, and T. Kubo, "Sign Language 'Heard' in the Auditory Cortex," *Nature* 397, no. 6715 (1999): 116.

9. E. I. Knudsen, G. G. Blasdel, and M. Konishi, "Sound Localization by the Barn Owl (Tyto-Alba) Measured with the Search Coil Technique," *Journal of Comparative Physiology* 133, no. 1 (1979): 1–11.

10. G. Ashida, "Barn Owl and Sound Localization," *Acoustical Science and Technology* 36, no. 4 (2015): 275–285.

11. E. I. Knudsen, "Instructed Learning in the Auditory Localization Pathway of the Barn Owl," *Nature* 417, no. 6886 (2002): 322–328.

12. M. S. Brainard and E. I. Knudsen, "Sensitive Periods for Visual Calibration of the Auditory Space Map in the Barn Owl Optic Tectum," *Journal of Neuroscience* 18, no. 10 (1998): 3929–3942.

13. B. A. Linkenhoker and E. I. Knudsen, "Incremental Training Increases the Plasticity of the Auditory Space Map in Adult Barn Owls," *Nature* 419, no. 6904 (2002): 293–296.

14. M. S. Brainard and E. I. Knudsen, "Sensitive Periods for Visual Calibration of the Auditory Space Map in the Barn Owl Optic Tectum," *Journal of Neuroscience* 18, no. 10 (1998): 3929–3942.

15. J. Fritz, S. Shamma, M. Elhilali, and D. Klein, "Rapid Task-Related Plasticity of Spectrotemporal Receptive Fields in Primary Auditory Cortex," *Nature Neuroscience* 6, no. 11 (2004): 1216–1223; M. Ahissar and S. Hochstein, "The Reverse Hierarchy Theory of Visual Perceptual Learning," *Trends in Cognitive Sciences* 8, no. 10 (2003): 457–464.

16. O. Kacelnik, F. R. Nodal, C. H. Parsons, and A. J. King, "Training-Induced Plasticity of Auditory Localization in Adult Mammals," *PloS Biology* 4, no. 4 (2006): e71.

17. V. M. Bajo, F. R. Nodal, D. R. Moore, and A. J. King, "The Descending Cortico-collicular Pathway Mediates Learning-Induced Auditory Plasticity," *Nature Neuroscience* 13, no. 2 (2010): 253–260.

18. A. H. Teich, P.M. McCabe, C. C. Gentile, L. S. Schneiderman, R. W. Winters, D. R. Liskowsky, and N. Schneiderman, "Auditory Cortex Lesions Prevent the Extinction of Pavlovian Differential Heart Rate Conditioning to Tonal Stimuli in Rabbits," *Brain Research* 480, nos. 1–2 (1989): 210–218.

19. X. F. Ma and N. Suga, "Plasticity of Bat's Central Auditory System Evoked by Focal Electric Stimulation of Auditory and/or Somatosensory Cortices," *Journal of Neurophysiology* 85, no. 3 (2001): 1078–1087.

20. Y. Zhang, N. Suga, and J. Yan, "Corticofugal Modulation of Frequency Processing in Bat Auditory System," *Nature* 387, no. 6636 (1997): 900–903.

21. N. Suga and X. F. Ma, "Multiparametric Corticofugal Modulation and Plasticity in the Auditory System," *Nature Reviews. Neuroscience* 4, no. 10 (2003): 783–794.

22. F. Luo, Q. Wang, A. Kashani, and J. Yan, "Corticofugal Modulation of Initial Sound Processing in the Brain," *Journal of Neuroscience* 28, no. 45 (2008): 11615–11621.

23. M. V. Popescu and D. B. Polley, "Monaural Deprivation Disrupts Development of Binaural Selectivity in Auditory Midbrain and Cortex," *Neuron* 65, no. 5 (2010): 718–731.

24. P. Dallos, B. Evans, and R. Hallworth, "Nature of the Motor Element in Electrokinetic Shape Changes of Cochlear Outer Hair Cells," *Nature* 350, no. 6314 (1991): 155–157.

25. P. J. Dallos, "On Generation of Odd-Fractional Subharmonics," *Journal of the Acoustical Society of America* 40, no. 6 (1966): 1381–1391; D. T. Kemp, "Stimulated Acoustic Emissions from within the Human Auditory System," *Journal of the Acoustical Society of America* 64, no. 5 (1978): 1386–1991.

26. M. C. Liberman, "The Olivocochlear Efferent Bundle and Susceptibility of the Inner Ear to Acoustic Injury," *Journal of Neurophysiology* 65, no. 1 (1991): 123–132.

27. X. Perrot, P. Ryvlin, J. Isnard, M. Guenot, H. Catenoix, C. Fischer, F. Mauguiere, and L. Collet, "Evidence for Corticofugal Modulation of Peripheral Auditory

Activity in Humans," *Cerebral Cortex* 16, no. 7 (2006)): 941–948; S. Khalfa, R. Bougeard, N. Morand, E. Veuillet, J. Isnard, M. Guenot, P. Ryvlin, C. Fischer, and L. Collet, "Evidence of Peripheral Auditory Activity Modulation by the Auditory Cortex in Humans," *Neuroscience* 104, no. 2 (2001): 347–358.

28. P. Froehlich, L. Collet, and A. Morgon, "Transiently Evoked Otoacoustic Emission Amplitudes Change with Changes of Directed Attention," *Physiology and Behavior* 53, no. 4 (1993): 679–682; C. Meric and L. Collet, "Differential Effects of Visual Attention on Spontaneous and Evoked Otoacoustic Emissions," *International Journal of Psychophysiology* 17, no. 3 (1994): 281–289; S. Srinivasan, A. Keil, K. Stratis, K. L. Woodruff Carr, and D. W. Smith, "Effects of Cross-Modal Selective Attention on the Sensory Periphery: Cochlear Sensitivity Is Altered by Selective Attention," *Neuroscience* 223 (2012): 325–332.

29. X. Perrot, C. Micheyl, S. Khalfa, and L. Collet, "Stronger Bilateral Efferent Influences on Cochlear Biomechanical Activity in Musicians Than in Non-Musicians," *Neuroscience Letters* 262, no. 3 (1999): 167–170; C. Micheyl, S. Khalfa, X. Perrot, and L. Collet, "Difference in Cochlear Efferent Activity between Musicians and Non-Musicians," *Neuroreport* 8, no. 4 (1997): 1047–50; S. M. Brashears, T. G. Morlet, C. I. Berlin, and L. J. Hood, "Olivocochlear Efferent Suppression in Classical Musicians," *Journal of the American Academy of Audiology* 14, no. 6 (2003): 314–324.

30. V. Marian, T. Q. Lam, S. Hayakawa, and S. Dhar, "Spontaneous Otoacoustic Emissions Reveal an Efficient Auditory Efferent Network," *Journal of Speech, Language, and Hearing Research* 61, no. 11 (2018): 2827–2832.

31. M. E. Goldberg and R. H. Wurtz, "Activity of Superior Colliculus in Behaving Monkey 2. Effect of Attention on Neuronal Responses," *Journal of Neurophysiology* 35, no. 4 (1972): 560–574.

32. C. G. Kentros, N. T. Agnihotri, S. Streater, R. D. Hawkins, and E. R. Kandel, "Increased Attention to Spatial Context Increases Both Place Field Stability and Spatial Memory," *Neuron* 42, no. 2 (2004): 283–295.

33. E. R. Kandel, *In Search of Memory: The Emergence of a New Science of Mind* (New York: W. W. Norton, 2006).

34. Quoted in Matt Richtel, "Outdoors and Out of Reach, Studying the Brain," *New York Times*, August 15, 2010, https://www.nytimes.com/2010/08/16/technology/16brain.html.

35. J. Fritz, S. Shamma, M. Elhilali, and D. Klein, "Rapid Task-Related Plasticity of Spectrotemporal Receptive Fields in Primary Auditory Cortex," *Nature Neuroscience* 6, no. 11 (2003): 1216–1223.

36. J. B. Fritz, M. Elhilali, and S. A. Shamma, "Differential Dynamic Plasticity of A1 Receptive Fields during Multiple Spectral Tasks," *Journal of Neuroscience* 25, no. 33 (2005): 7623–7635.

37. J. Fritz, M. Elhilali, and S. Shamma, "Active Listening: Task-Dependent Plasticity of Spectrotemporal Receptive Fields in Primary Auditory Cortex," *Hearing Research* 206, no. 1–2 (2005): 159–176.

38. S. J. Slee and S. V. David, "Rapid Task-Related Plasticity of Spectrotemporal Receptive Fields in the Auditory Midbrain," *Journal of Neuroscience* 35, no. 38 (2015): 13090–13102.

39. P. H. Delano, D. Elgueda, C. M. Hamame, and L. Robles, "Selective Attention to Visual Stimuli Reduces Cochlear Sensitivity in Chinchillas," *Journal of Neuroscience* 27, no. 15 (2007): 4146–4153.

40. N. Mesgarani and E. F. Chang, "Selective Cortical Representation of Attended Speaker in Multi-Talker Speech Perception," *Nature* 485, no. 7397 (2012): 233–236; J. Krizman, A. Tierney, T. Nicol, and N. Kraus, "Attention Induces a Processing Tradeoff between Midbrain and Cortex," in *Association for Research in Otolaryngology* PS 428 (2017): 277.

41. N. M. Weinberger, A. A. Miasnikov, and J. C. Chen, "The Level of Cholinergic Nucleus Basalis Activation Controls the Specificity of Auditory Associative Memory," *Neurobiology of Learning and Memory* 86 (2006): 270–285.

42. H. H. Webster, U. K. Hanisch, R. W. Dykes, and D. Biesold, "Basal Forebrain Lesions with or without Reserpine Injection Inhibit Cortical Reorganization in rat Hindpaw Primary Somatosensory Cortex Following Sciatic Nerve Section," *Somatosensory & Motor Research* 8 (1991): 327–346.

43. M. P. Kilgard and M. M. Merzenich, "Cortical Map Reorganization Enabled by Nucleus Basalis Activity," *Science* 279 (1998): 1714–1718.

44. W. Guo, B. Robert, and D. B. Polley, "The Cholinergic Basal Forebrain Links Auditory Stimuli with Delayed Reinforcement to Support Learning," *Neuron* 103, no. 6 (2019): P1164–1177.E6.

45. S. Corkin, "Acquisition of Motor Skill After Bilateral Medial Temporal-Lobe Excision," *Neuropsychologia* 6, no. 3 (1968): 255–265.

46. J. R. Saffran, R. N. Aslin, and E. L. Newport, "Statistical Learning by 8-Month-Old Infants," *Science* 274, no. 5294 (1996): 1926–1928; E. Partanen, T. Kujala, R. Näätänen, A. Liitola, A. Sambeth, and M. Huotilainen, "Learning-Induced Neural Plasticity of Speech Processing Before Birth," *Proceedings of the National Academy of Sciences* 110, no. 37 (2013): 15145–15150.

47. J. Fritz, S. Shamma, M. Elhilali, and D. Klein, "Rapid Task-Related Plasticity of Spectrotemporal Receptive Fields in Primary Auditory Cortex," *Nature Neuroscience* 6, no. 11 (2003): 1216–1223.

48. N. Kraus and T. White-Schwoch, "Unraveling the Biology of Auditory Learning: A Cognitive-Sensorimotor-Reward Framework," *Trends in Cognitive Sciences* 19 (2015): 642–654.

Chapter 4

1. I. Fried, K. A. MacDonald, and C. L. Wilson, "Single Neuron Activity in Human Hippocampus and Amygdala During Recognition of Faces and Objects," *Neuron* 18, no. 5 (1997): 753–765.

2. J. B. Meixner and J. P. Rosenfeld, "Detecting Knowledge of Incidentally Acquired, Real-World Memories Using a P300-Based Concealed-Information Test," *Psychological Science* 25, no. 11 (2014): 1994–2005; J. B. Meixner and J. P. Rosenfeld, "A Mock Terrorism Application of the P300-Based Concealed Information Test," *Psychophysiology* 48, no. 2 (2011): 149–154.

3. R. Näätänen, *Attention and Brain Function* (Hillsdale, NJ: Erlbaum, 1992).

4. R. Näätänen, A. W. Gaillard, and S. Mäntysalo, "Early Selective-Attention Effect on Evoked Potential Reinterpreted," *Acta Psychologica* 42, no. 4 (1978): 313–329.

5. M. Sams, P. Paavilainen, K. Alho, and R. Näätänen, "Auditory Frequency Discrimination and Event-Related Potentials," *Electroencephalography and Clinical Neurophysiology* 62, no. 6 (1985): 437–448.

6. J. Allen, N. Kraus, and A. R. Bradlow, "Neural Representation of Consciously Imperceptible Speech-Sound Differences," *Perception and Psychophysics* 62 (2000): 1383–1393.

7. K. Tremblay, N. Kraus, and T. McGee, "The Time Course of Auditory Perceptual Learning: Neurophysiological Changes During Speech-Sound Training," *Neuroreport* 9, no. 16 (1998): 3557–3560.

8. T. McGee, N. Kraus, and T. Nicol, "Is It Really a Mismatch Negativity? An Assessment of Methods for Determining Response Validity in Individual Subjects," *Electroencephalography and Clinical Neurophysiology* 104, no. 4 (1997): 359–368.

9. F. G. Worden and J. T. Marsh, "Frequency-Following (Microphonic-Like) Neural Responses Evoked by Sound," *Electroencephalography and Clinical Neurophysiology* 25, no. 1 (1968): 42–52.

10. G. C. Galbraith, P. W. Arbagey, R. Branski, N. Comerci, and P. M. Rector, "Intelligible Speech Encoded in the Human Brain Stem Frequency-Following Response," *Neuroreport* 6, no. 17 (1995): 2363–2367; G. C. Galbraith, S. P. Jhaveri, and J. Kuo, "Speech-Evoked Brainstem Frequency-Following Responses During Verbal Transformations Due to Word Repetition," *Electroencephalography and Clinical Neurophysiology*

102, no. 1 (1997): 46–53; G. C. Galbraith, S. M. Bhuta, A. K. Choate, J. M. Kitahara, and T. A. Mullen, "Brain Stem Frequency-Following Response to Dichotic Vowels During Attention," *Neuroreport* 9, no. 8 (1998): 1889–1893.

11. A. Krishnan, Y. S. Xu, J. Gandour, and P. Cariani, "Encoding of Pitch in the Human Brainstem Is Sensitive to Language Experience," *Brain Research. Cognitive Brain Research* 25, no. 1 (2005): 161–168.

12. E. Skoe and N. Kraus, "Auditory Brainstem Response to Complex Sounds: A Tutorial," *Ear and Hearing* 31, no. 3 (2010): 302–24; J. Krizman and N. Kraus, "Analyzing the FFR: A Tutorial for Decoding the Richness of Auditory Function," *Hearing Research* 382 (2019): 107779; N. Kraus & T. Nicol "The Power of Sound for Brain Health," Nature Human Behaviour 1 (2017): 700-702.

13. J. Feldman, "The Neural Binding Problem(s)," *Cognitive Neurodynamics* 7, no. 1 (2013): 1–11.

14. I. McGilchrist, *The Master and His Emissary* (New Haven: Yale University Press, 2009).

15. J. Panksepp, *Affective Neuroscience: The Foundations of Human and Animal Emotions* (New York: Oxford University Press, 1998).

16. E. Coffey, T. Nicol, T. White-Schwoch, B. Chandrasekaran, J. Krizman, E. Skoe, R. Zatorre, and N. Kraus, "Evolving Perspectives on the Sources of the Frequency-Following Response," *Nature Communications* 10 (2019): 5036; L. Selinger, K. Zarnowiec, M. Via, I. C. Clemente, and C. Escera, "Involvement of the Serotonin Transporter Gene in Accurate Subcortical Speech Encoding," *Journal of Neuroscience* 36, no. 42 (2016): 10782–10790.

17. N. Kraus and T. White-Schwoch, "Unraveling the Biology of Auditory Learning: A Cognitive-Sensorimotor-Reward Framework," *Trends in Cognitive Sciences* 19 (2015): 642–654.

Chapter 5

1. E. A. Spitzka, "A Study of the Brains of Six Eminent Scientists and Scholars Belonging to the American Anthropometric Society, Together with a Description of the Skull of Professor E. D. Cope," *Transactions of the American Philosophical Society* 21, no. 4 (1907): 175–308.

2. J. Brandt, *The Grape Cure* (New York: The Order of Harmony, 1928).

3. S. Auerbach, "Zur Lokalisation des musicalischen Talentes im Gehirn unad am Schädel," *Archives of Anatomy and Physiology* (1906): 197–230.

4. P. Schneider, M. Scherg, H. G. Dosch, H. J. Specht, A. Gutschalk, and A. Rupp, "Morphology of Heschl's Gyrus Reflects Enhanced Activation in the Auditory Cortex of Musicians," *Nature Neuroscience* 5, no. 7 (2002): 688–694.

5. T. Elbert, C. Pantev, C. Wienbruch, B. Rockstroh, and E. Taub, "Increased Cortical Representation of the Fingers of the Left Hand in String Players," *Science* 270, no. 5234 (1995): 305–307.

6. G. Schlaug, "The Brain of Musicians: A Model for Functional and Structural Adaptation," *Annals of the New York Academy of Sciences* 930 (2001): 281–299.

7. D. J. Lee, Y. Chen, and G. Schlaug, "Corpus Callosum: Musician and Gender Effects," *Neuroreport* 14, no. 2 (2003): 205–209; G. Schlaug, L. Jäncke, Y. X. Huang, J. F. Staiger, and H. Steinmetz, "Increased Corpus-Callosum Size in Musicians," *Neuropsychologia* 33, no. 8 (1995): 1047.

8. S. Hutchinson, L. H. L. Lee, N. Gaab, and G. Schlaug, "Cerebellar Volume of Musicians," *Cerebral Cortex* 13, no. 9 (2003): 943–949.

9. F. Bouhali, V. Mongelli, M. Thiebaut, and L. Cohen, "Reading Music and Words: The Anatomical Connectivity of Musicians' Visual Cortex," *Neuroimage* 212 (2020): 116666.

10. S. L. Bengtsson, Z. Nagy, S. Skare, L. Forsman, H. Forssberg, and F. Ullen, "Extensive Piano Practicing Has Regionally Specific Effects on White Matter Development," *Nature Neuroscience* 8, no. 9 (2005): 1148–1150.

11. C. Pantev, R. Oostenveld, A. Engelien, B. Ross, L. E. Roberts, and M. Hoke, "Increased Auditory Cortical Representation in Musicians," *Nature* 392, no. 6678 (1998): 811–814; A. Shahin, L. E. Roberts, and L. J. Trainor, "Enhancement of Auditory Cortical Development by Musical Experience in Children," *Neuroreport* 15, no. 12 (2004): 1917–21; A. J. Shahin, L. E. Roberts, W. Chau, L. J. Trainor, and L. M. Miller, "Music Training Leads to the Development of Timbre-Specific Gamma Band Activity," *Neuroimage* 41, no. 1 (2008): 113–122; A. Shahin, D. J. Bosnyak, L. J. Trainor, and L. E. Roberts, "Enhancement of Neuroplastic P2 and N1c Auditory Evoked Potentials in Musicians," *Journal of Neuroscience* 23, no. 13 (1998): 5545–5552.

12. S. Koelsch, E. Schroger, and M. Tervaniemi, "Superior Pre-Attentive Auditory Processing in Musicians," *Neuroreport* 10, no. 6 (1999): 1309–1313; E. Brattico, K. J. Pallesen, O. Varyagina, C. Bailey, I. Anourova, M. Jarvenpaa, T. Eerola, and M. Tervaniemi, "Neural Discrimination of Nonprototypical Chords in Music Experts and Laymen: An MEG Study," *Journal of Cognitive Neuroscience* 21, no. 11 (2009): 2230–2244.

13. P. Virtala, M. Huotilainen, E. Lilja, J. Ojala, and M. Tervaniemi, "Distortion and Western Music Chord Processing: An ERP Study of Musicians and Nonmusicians," *Music Perception* 35, no. 3 (2018): 315–331.

14. A. Parbery-Clark, S. Anderson, E. Hittner, and N. Kraus, "Musical Experience Strengthens the Neural Representation of Sounds Important for Communication in Middle-Aged Adults," *Frontiers in Aging Neuroscience* 4, no. 30 (2012): 1–12; N. Kraus and B. Chandrasekaran, "Music Training for the Development of Auditory Skills,"

Nature Reviews Neuroscience 11 (2010): 599–605; N. Kraus and T. White-Schwoch, "Neurobiology of Everyday Communication: What Have We Learned from Music?" *Neuroscientist* 23, no. 3 (2017): 287–298.

15. N. Kraus and T. White-Schwoch, "Unraveling the Biology of Auditory Learning: A Cognitive-Sensorimotor-Reward Framework," *Trends in Cognitive Sciences* 19 (2015): 642–654.

16. M. Tervaniemi, L. Janhunen, S. Kruck, V. Putkinen, and M. Huotilainen, "Auditory Profiles of Classical, Jazz, and Rock Musicians: Genre-Specific Sensitivity to Musical Sound Features," *Frontiers in Psychology* 6 (2015): 1900.

17. M. Tervaniemi, M. Rytkonen, E. Schroger, R. J. Ilmoniemi, and R. Naatanen, "Superior Formation of Cortical Memory Traces for Melodic Patterns in Musicians," *Learning and Memory* 8, no. 5 (2001): 295–300.

18. E. Brattico, K. J. Pallesen, O. Varyagina, C. Bailey, I. Anourova, M. Jarvenpaa, T. Eerola, and M. Tervaniemi, "Neural Discrimination of Nonprototypical Chords in Music Experts and Laymen: An MEG Study," *Journal of Cognitive Neuroscience* 21, no. 11 (2009): 2230–2244; S. Leino, E. Brattico, M. Tervaniemi, and P. Vuust. "Representation of Harmony Rules in the Human Brain: Further Evidence from Event-Related Potentials," *Brain Research* 1142 (2007): 169–177; P. Virtala, M. Huotilainen, E. Partanen, and M. Tervaniemi, "Musicianship Facilitates the Processing of Western Music Chords—an ERP and Behavioral Study," *Neuropsychologia* 61 (2014): 247–258; W. De Baene, A. Vandierendonck, M. Leman, A. Widmann, and M. Tervaniemi, "Roughness Perception in Sounds: Behavioral and ERP Evidence," *Biological Psychology* 67, no. 3 (2004): 319–330; M. Tervaniemi, V. Just, S. Koelsch, A. Widmann, and E. Schroger, "Pitch Discrimination Accuracy in Musicians vs Nonmusicians: An Event-Related Potential and Behavioral Study," *Experimental Brain Research* 161, no. 1 (2005): 1–10; M. Tervaniemi, E. Huotilainen, E. Brattico, R. J. Ilmoniemi, K. Reinikainen, and K. Alho, "Event-Related Potentials to Expectancy Violation in Musical Context," *Musicae Scientiae* 7, no. 2 (2003): 241–261; A. Caclin, E. Brattico, B. K. Smith, M. Ternaviemi, M.-H. Giard, and S. McAdams, "Electrophysiological Correlates of Musical Timbre Perception," *Journal of the Acoustical Society of America* 112, no. 5 (2002): 2240; M. Tervaniemi, A. Castaneda, M. Knoll, and M. Uther, "Sound Processing in Amateur Musicians and Nonmusicians: Event-Related Potential and Behavioral Indices," *Neuroreport* 17, no. 11 (2006): 1225–1258.

19. A. Parbery-Clark, S. Anderson, E. Hittner, and N. Kraus, "Musical Experience Strengthens the Neural Representation of Sounds Important for Communication in Middle-Aged Adults," *Frontiers in Aging Neuroscience* 4, no. 30 (2012): 1–12; N. Kraus and B. Chandrasekaran, "Music Training for the Development of Auditory Skills," *Nature Reviews Neuroscience* 11 (2010): 599–605; N. Kraus and T. White-Schwoch, "Neurobiology of Everyday Communication: What Have We Learned from Music?

Neuroscientist 23, no. 3 (2017): 287–298; D. L. Strait, A. Parbery-Clark, E. Hittner, and N Kraus, "Musical Training During Early Childhood Enhances the Neural Encoding of Speech in Noise," *Brain and Language* 123, no. 3 (2012): 191–201; D. L. Strait, A. Parbery-Clark, S. O'Connell, and N. Kraus, "Biological Impact of Preschool Music Classes on Processing Speech in Noise," *Developmental Cognitive Neuroscience* 6 (2013): 51–60; A. Parbery-Clark, E. Skoe, and N. Kraus, "Musical Experience Limits the Degradative Effects of Background Noise on the Neural Processing of Sound," *Journal of Neuroscience* 29, no. 45 (2009): 14100–14107.

20. C. Pantev, L. E. Roberts, M. Schulz, A. Engelien, and B. Ross, "Timbre-Specific Enhancement of Auditory Cortical Representations in Musicians," *Neuroreport* 12, no. 1 (2001): 169–174.

21. E. H. Margulis, L. M. Mlsna, A. K. Uppunda, T. B. Parrish, and P. C. M. Wong, "Selective Neurophysiologic Responses to Music in Instrumentalists with Different Listening Biographies," *Human Brain Mapping* 30, no. 1 (2009): 267–275.

22. D. L. Strait, K. Chan, R. Ashley, and N. Kraus, "Specialization among the Specialized: Auditory Brainstem Function Is Tuned in to Timbre," *Cortex* 48 (2012): 360–362.

23. T. F. Münte, C. Kohlmetz, W. Nager, and E. Altenmüller, "Superior Auditory Spatial Tuning in Conductors," *Nature* 409, no. 6820 (2001): 580.

24. N. Matthews, L. Welch, and E. Festa, "Superior Visual Timing Sensitivity in Auditory but Not Visual World Class Drum Corps Experts," *eNeuro* 5, no. 6 (2018).

25. G. Musacchia, M. Sams, E. Skoe, and N. Kraus, "Musicians Have Enhanced Subcortical Auditory and Audiovisual Processing of Speech and Music," *Proceedings of the National Academy of Sciences of the United States of America* 104, no. 40 (2007): 15894–15898.

26. J. L. Chen, V. B. Penhune, and R. J. Zatorre, "Listening to Musical Rhythms Recruits Motor Regions of the Brain," *Cerebral Cortex* 18, no. 12 (2008): 2844–54; A. Lahav, E. Saltzman, and G. Schlaug, "Action Representation of Sound: Audiomotor Recognition Network while Listening to Newly Acquired Actions," *Journal of Neuroscience* 27, no. 2 (2007): 308–314.

27. F. J. Langheim, J. H. Callicott, V. S. Mattay, J. H. Duyn, and D. R. Weinberger, "Cortical Systems Associated with Covert Music Rehearsal," *Neuroimage* 16, no. 4 (2002): 901–908; A. R. Halpern and R. J. Zatorre, "When That Tune Runs through Your Head: A PET Investigation of Auditory Imagery for Familiar Melodies," *Cerebral Cortex* 9, no. 7 (1999): 697–704.

28. K. Amunts, G. Schlaug, A. Schleicher, H. Steinmetz, A. Dabringhaus, P. E. Roland, and K. Zilles, "Asymmetry in the Human Motor Cortex and Handedness," *Neuroimage* 4, no. 3 part 1 (1996): 216–222; L. E. White, G. Lucas, A. Richards, and D. Purves, "Cerebral Asymmetry and Handedness," *Nature* 368, no. 6468 (1994): 197–198.

29. C. Gaser and G. Schlaug, "Gray Matter Differences between Musicians and Non-musicians," *Annals of the New York Academy of Sciences* 999 (2003): 514–517.

30. T. Elbert, C. Pantev, C. Wienbruch, B. Rockstroh, and E. Taub, "Increased Cortical Representation of the Fingers of the Left Hand in String Players," *Science* 270, no. 5234 (1995): 305–307.

31. H. Corrigall and E. G. Schellenberg, "Music: The Language of Emotion," in *Handbook of Psychology of Emotions,* ed. C. Mohiyeddini, M. Eyesenck, and S. Bauer (Hauppauge, NY: Nova Science Publishers, 2013), 299–326.

32. M. Iwanaga and Y. Moroki, "Subjective and Physiological Responses to Music Stimuli Controlled over Activity and Preference," *Journal of Music Therapy* 36, no. 1 (1999): 26–38; L.-O. Lundqvist, F. Carlsson, P. Hilmersson, and P. N. Juslin, "Emotional Responses to Music: Experience, Expression, and Physiology," *Psychology of Music* 37, no. 1 (2009): 61–90; R. A. McFarland, "Relationship of Skin Temperature Changes to the Emotions Accompanying Music," *Biofeedback and Self-Regulation* 10 (1985): 255–267; C. L. Krumhansl, "An Exploratory Study of Musical Emotions and Psychophysiology," *Canadian Journal of Experimental Psychology* 51, no. 4 (1997): 336–353.

33. H. Corrigall and E. G. Schellenberg, "Music: The Language of Emotion," in *Handbook of Psychology of Emotions,* ed. C. Mohiyeddini, M. Eyesenck, and S. Bauer (Hauppauge, NY: Nova Science Publishers, 2013), 299–326.

34. A. J. Blood and R. J. Zatorre, "Intensely Pleasurable Responses to Music Correlate with Activity in Brain Regions Implicated in Reward and Emotion," *Proceedings of the National Academy of Sciences of the United States of America* 98, no. 20 (2001): 11818–11823.

35. V. N. Salimpoor, M. Benovoy, K. Larcher, A. Dagher, and R. J. Zatorre, "Anatomically Distinct Dopamine Release During Anticipation and Experience of Peak Emotion to Music," *Nature Neuroscience* 14, no. 2 (2011): 257–256.

36. V. N. Salimpoor, I. van den Bosch, N. Kovacevic, A. R. McIntosh, A. Dagher, and R. J. Zatorre, "Interactions between the Nucleus Accumbens and Auditory Cortices Predict Music Reward Value," *Science* 340, no. 6129 (2013): 216–219.

37. E. Mas-Herrero, R. J. Zatorre, A. Rodriguez-Fornells, and J. Marco-Pallares, "Dissociation between Musical and Monetary Reward Responses in Specific Musical Anhedonia," *Current Biology* 24, no. 6 (2014): 699–704.

38. N. Martinez-Molina, E. Mas-Herrero, A. Rodriguez-Fornells, R. J. Zatorre, and J. Marco-Pallares, "Neural correlates of specific musical anhedonia," *Proceedings of the National Academy of Sciences of the United States of America* 113, no. 46 (2016): E7337–345.

39. D. Strait, E. Skoe, N. Kraus, and R. Ashley, "Musical Experience and Neural Efficiency: Effects of Training on Subcortical Processing of Vocal Expressions of Emotion," *European Journal of Neuroscience* 29 (2009): 661–668.

40. A. S. Chan, Y. C. Ho, and M. C. Cheung, "Music Training Improves Verbal Memory," *Nature* 396, no. 6707 (1998): 128; Y. C. Ho, M. C. Cheung, and A. S. Chan, "Music Training Improves Verbal but Not Visual Memory: Cross-Sectional and Longitudinal Explorations in Children," *Neuropsychology* 17, no. 3 (2003): 439–450; L. S. Jakobson, S. T. Lewycky, A. R. Kilgour, and B. M. Stoesz, "Memory for Verbal and Visual Material in Highly Trained Musicians," *Music Perception* 26, no. 1 (2008): 41–55; A. T. Tierney, T. R. Bergeson-Dana, and D. B. Pisoni, "Effects of Early Musical Experience on Auditory Sequence Memory," *Empirical Musicology Revirew* 3, no. 4 (2008): 178–186; S. Brandler and T. H. Rammsayer, "Differences in Mental Abilities between Musicians and Non-Musicians," *Psychology of Music* 31, no. 2 (2003): 123–138; M. S. Franklin, K. S. Moore, K. Rattray, and J. Moher, "The Effects of Musical Training on Verbal Memory," *Psychology of Music* 36, no. 3 (2008): 353–365.

41. D. L. Strait, A. Parbery-Clark, S. O'Connell, and N. Kraus, "Biological Impact of Preschool Music Classes on Processing Speech in Noise," *Developmental Cognitive Neuroscience* 6 (2013): 51–60; A. Parbery-Clark, E. Skoe, and N. Kraus, "Musical Experience Limits the Degradative Effects of Background Noise on the Neural Processing of Sound," *Journal of Neuroscience* 29, no. 45 (2009): 14100–14107; A. Parbery-Clark, D. L. Strait, S. Anderson, E. Hittner, and N. Kraus, "Musical Experience and the Aging Auditory System: Implications for Cognitive Abilities and Hearing Speech in Noise," *PLOS ONE* 6, no. 5 (2011): E18082l; K. J. Pallesen, E. Brattico, C. J. Bailey, A. Korvenoja, J. Koivisto, A. Gjedde, and S. Carlson, "Cognitive Control in Auditory Working Memory Is Enhanced in Musicians," *PLOS ONE* 5, no. 6 (2010): E11120; D. Strait, S. O'Connell, A. Parbery-Clark, and N. Kraus, "Musicians' Enhanced Neural Differentiation of Speech Sounds Arises Early in Life: Developmental Evidence from Ages Three to Thirty," *Cerebral Cortex* 24, no. 9 (2014): 2512–2521; E. M. George and D. Coch, "Music Training and Working Memory: An ERP Study," *Neuropsychologia* 49, no. 5 (2011): 1083–1094; S. B. Nutley, F. Darki, and T. Klingberg. "Music Practice Is Associated with Development of Working Memory During Childhood and Adolescence," *Frontiers in Human Neuroscience* (2014); G. M. Bidelman, S. Hutka, and S. Moreno, "Tone Language Speakers and Musicians Share Enhanced Perceptual and Cognitive Abilities for Musical Pitch: Evidence for Bidirectionality between the Domains of Language and Music," *PLOS ONE* 8, no. 4 (2013): E60676.

42. A. T. Tierney, T. R. Bergeson-Dana, and D. B. Pisoni, "Effects of Early Musical Experience on Auditory Sequence Memory," *Empirical Musicology Review* 3, no. 4 (2007): 178–186; Y. Lee, M. Lu, and H. Ko, "Effects of Skill Training on Working Memory Capacity," *Learning and Instruction* 17, no. 3 (2007): 336–344.

43. J. Zuk, C. Benjamin, A. Kenyon, and N. Gaab, "Behavioral and Neural Correlates of Executive Functioning in Musicians and Non-Musicians," *PLOS ONE* 9, no. 6 (2014): E99868; L. Moradzadeh, G. Blumenthal, and M. Wiseheart, "Musical Training, Bilingualism, and Executive Function: A Closer Look at Task Switching and Dual-Task Performance," *Cognitive Sciences* 39, no. 5 (2015): 992–1020; A. C.

Jaschke, H. Honing, and E. J. A. Scherder, "Longitudinal Analysis of Music Educa-tion on Executive Functions in Primary School Children," *Frontiers in Neuroscience* (2018): 12; E. Bialystok and A. M. Depape, "Musical Expertise, Bilingualism, and Executive Functioning," *Journal of Experimental Psychology: Human Perception and Performance* 35, no. 2 (2009): 565–574; D. Strait, N. Kraus, A. Parbery-Clark, and R. Ashley, "Musical Experience Shapes Top-Down Auditory Mechanisms: Evidence from Masking and Auditory Attention Performance," *Hearing Research* 261 (2010): 22–29; K. K. Clayton, J. Swaminathan, A. Yazdanbakhsh, J. Zuk, A. D. Patel, and G. Kidd Jr., "Executive Function, Visual Attention and the Cocktail Party Problem in Musicians and Non-Musicians," *PLoS One* 11, no. 7 (2016): E0157638; A. J. Oxen-ham, B. J. Fligor, C. R. Mason and G. Kidd, "Informational Masking and Musical Training," *Journal of the Acoustical Society of America* 114, no. 3 (2003): 1543–1549.

44. K. J. Pallesen, E. Brattico, C. J. Bailey, A. Korvenoja, J. Koivisto, A. Gjedde, and S. Carlson, "Cognitive Control in Auditory Working Memory Is Enhanced in Musi-cians," *PLOS ONE* 5, no. 6 (2010): e11120; J. Zuk, C. Benjamin, A. Kenyon, and N. Gaab, "Behavioral and Neural Correlates of Executive Functioning in Musicians and Non-Musicians," *PLOS ONE* 9, no. 6 (2014): e99868; K. Schulze, K. Mueller, and S. Koelsch, "Neural Correlates of Strategy Use During Auditory Working Memory in Musicians and Non-Musicians," *European Journal of Neuroscience* 33, no. 1 (2011): 189–196; K. Schulze, S. Zysset, K. Mueller, A. D. Friederici, and S. Koelsch, "Neuroar-chitecture of Verbal and Tonal Working Memory in Nonmusicians and Musicians," *Human Brain Mapping* 32, no. 5 (2011): 771–783.

45. D. L. Strait, K. Chan, R. Ashley, and N. Kraus, "Specialization Among the Spe-cialized: Auditory Brainstem Function Is Tuned in to Timbre," *Cortex* 48 (2012): 360–362; N. Kraus, D. Strait, and A. Parbery-Clark, "Cognitive Factors Shape Brain Networks for Auditory Skills: Spotlight on Auditory Working Memory," *Annals of the New York Academy of Sciences* 1252 (2012): 100–107; D. L. Strait, J. Hornickel, and N. Kraus, "Subcortical Processing of Speech Regularities Underlies Reading and Music Aptitude in Children," *Behavioral and Brain Functions* 7, no. 1 (2011): 44; D. L. Strait, S. O'Connell, A. Parbery-Clark, and N. Kraus, "Musicians' Enhanced Neural Differentiation of Speech Sounds Arises Early in Life: Developmental Evidence from Ages 3 to 30," *Cerebral Cortex* (2013): https:doi.org/10.1093/cercor/bht103.

46. C. J. Limband and A. R. Braun, "Neural Substrates of Spontaneous Musical Per-formance: An FMRI Study of Jazz Improvisation," *PLOS ONE* 3, no. 2 (2008): e1679.

47. J. Collier, "Musician Explains One Concept in 5 Levels of Difficulty," *Wired*, YouTube video, January 8, 2018, https://www.youtube.com/watch?v=eRkgK4jfi6M.

48. T. Gioia, *Healing Songs* (Durham, NC: Duke University Press, 2006).

49. S. Bodeck, C. Lappe, and S. Evers, "Tic-Reducing Effects of Music in Patients with Tourette's Syndrome: Self-Reported and Objective Analysis," *Journal of the Neu-rological Sciences* 352, no. 1–2 (2015): 41–47.

50. O. Sacks, *Musicophilia: Tales of Music and the Brain* (New York: Alfred A. Knopf, 2007).

51. T. Gioia, *Healing Songs* (Durham, NC: Duke University Press, 2006).

52. C. M. Tomaino, "Clinical Applications of Music Therapy in Neurologic Rehabilitation," in *Music That Works*, ed. R. B. Haas and V. Brandes (Austria: Springer-Verlag, 2009), 211–220.

53. S. Hegde, "Music-Based Cognitive Remediation Therapy for Patients with Traumatic Brain Injury," *Frontiers in Neurology* 5 (2014): 34; M. H. Thaut, J. C. Gardiner, D. Holmberg, J. Horwitz, L. Kent, G. Andrews, B. Donelan, and G. R. McIntosh, "Neurologic Music Therapy Improves Executive Function and Emotional Adjustment in Traumatic Brain Injury Rehabilitation," *Annals of the New York Academy of Sciences* 1169 (2009): 406–416.

54. K. Bergmann, "The Sound of Trauma: Music Therapy in a Post-War Environment," *Australian Journal of Music Therapy* 13 (2012): 3–16; M. Bensimon, D. Amir, and Y. Wolf, "Drumming Through Trauma: Music Therapy with Post-Traumatic Soldiers," *Arts in Psychotherapy* 35, no. 1 (2008): 34–48; S. Garrido, F. A. Baker, J. W. Davidson, G. Moore, and S. Wasserman, "Music and Trauma: The Relationship between Music, Personality, and Coping Style," *Frontiers in Psychology* 6 (2015): 977; J. Loewy and K. Stewart, "Music Therapy to Help Traumatized Children and Caretakers," in *Mass Trauma and Violence*, ed. N. B. Webb, 191–215 (New York: Guilford Press, 2004); J. V. Loewy and A. F. Hara, *Caring for the Caregiver: The Use of Music Therapy in Grief and Trauma* (The American Music Therapy Association, 2002); J. Orth, L. Doorschodt, J. Verburgt, and B. Drožđek, "Sounds of Trauma: An Introduction to Methodology in Music Therapy with Traumatized Refugees in Clinical and Outpatient Settings," in *Broken Spirits: The Treatment of Traumatized Asylum Seekers, Refugees, War, and Torture Victims*, ed. J. Willson and B. Drožđek, 443–80 (New York: Brunner-Routledge, 2004).

55. S. L. Robb, D. S. Burns, K. A. Stegenga, P. R. Haut, P. O. Monahan, J. Meza, T. E. Stump, et al., "Randomized Clinical Trial of Therapeutic Music Video Intervention for Resilience Outcomes in Adolescents/Young Adults Undergoing Hematopoietic Stem Cell Transplant," *Cancer* 120, no. 6 (2014): 909–917.

56. C. M. Tomaino, "Meeting the Complex Needs of Individuals with Dementia through Music Therapy," *Music and Medicine* 5, no. 4 (2013): 234–241.

57. M. W. Hardy and A. B. Lagasse, "Rhythm, Movement, and Autism: Using Rhythmic Rehabilitation Research As a Model for Autism," *Frontiers in Integrative Neuroscience* 7 (2013): 19; A. B. LaGasse, "Effects of a Music Therapy Group Intervention on Enhancing Social Skills in Children with Autism," *Journal of Music Therapy* 51, no. 3 (2014): 250–275; A. B. LaGasse, "Social Outcomes in Children with Autism Spectrum Disorder: A Review of Music Therapy Outcomes," *Patient Related Outcome Measures* 8 (2017): 23–32.

58. W. Groß, U. Linden W, and T. Ostermann, "Effects of Music Therapy in the Treatment of Children with Delayed Speech Development—Results of a Pilot Study," *BMC Complementary and Alternative Medicine* 10 (2010): 39; M. Ritter, K. A. Colson, and J. Park, "Reading Intervention Using Interactive Metronome in Children with Language and Reading Impairment: A Preliminary Investigation," *Communication Disorders Quarterly* 34, no. 2 (2012): 106–119; G. E. Taub, K. S. McGrew, and T. Z. Keith, "Improvements in Interval Time Tracking and Effects on Reading Achievement," *Psychology in the Schools* 44, no. 8 (2007): 849–863.

59. C. Nombela, L. E. Hughes, A. M. Owen and J. A. Grahn, "Into the Groove: Can Rhythm Influence Parkinson's Disease?" *Neuroscience & Biobehavorial Reviews* 37, no. 10, pt. 2 (2013): 2564–2570; M. J. de Dreu, A. S. van der Wilk, E. Poppe, G. Kwakkel, and E. E. van Wegen, "Rehabilitation, Exercise Therapy and Music in Patients with Parkinson's Disease: A Meta-Analysis of the Effects of Music-Based Movement Therapy on Walking Ability, Balance and Quality of Life," *Parkinsonism & Related Disorders* 18 Suppl 1 (2012): S114–119; J. M. Hausdorff, J. Lowenthal, T. Herman, L. Gruendlinger, C. Peretz, and N. Giladi, "Rhythmic Auditory Stimulation Modulates Gait Variability in Parkinson's Disease," *European Journal of Neuroscience* 26, no. 8 (2007): 2369–2375; R. S. Calabro, A. Naro, S. Filoni, M. Pullia, L. Billeri, P. Tomasello, S. Portaro, G. Di Lorenzo, C. Tomaino, and P. Bramanti, "Walking to Your Right Music: A Randomized Controlled Trial on the Novel Use of Treadmill Plus Music in Parkinson's Disease," *Journal of Neuroengineering and Rehabilitation* 16, no. 1 (2019): 68.

60. A. Raglio, O. Oasi, M. Gianotti, A. Rossi, K. Goulene, and M. Stramba-Badiale, "Improvement of Spontaneous Language in Stroke Patients with Chronic Aphasia Treated with Music Therapy: A Randomized Controlled Trial," *Internal Journal of Neuroscience* 126, no. 3 (2016): 235–242; M. H. Thaut and G. C. McIntosh, "Neurologic Music Therapy in Stroke Rehabilitation," *Current Physical Medicine and Rehabilitation Reports* 2, no. 2 (2014): 106–113; J. P. Brady, "Metronome-Conditioned Speech Retraining for Stuttering," *Behavior Therapy* 2, no. 2 (1971): 129–150.

61. C. M. Tomaino, "Recovery of Fluent Speech through a Musician's Use of Prelearned Song Repertoire: A Case Study," *Music and Medicine* 2, no. (2010): 85–88; C. M. Tomaino, "Effective Music Therapy Techniques in the Treatment of Nonfluent Aphasia," *Annals of the New York Academy of Sciences* 1252, no. 1 (2012): 312–317; E. L. Stegemoller, T. R. Hurt, M. C. O'Connor, R. D. Camp, C. W. Green, J. C. Pattee, and E. K. Williams, "Experiences of Persons with Parkinson's Disease Engaged in Group Therapeutic Singing," *Journal of Music Therapy* 54, no. 4 (2018): 405–431.

62. A. Good, K. Gordon, B. C. Papsin, G. Nespoli, T. Hopyan, I. Peretz, and F. A. Russo, "Benefits of Music Training for Perception of Emotional Speech Prosody in Deaf Children with Cochlear Implants," *Ear and Hearing* 38, no. 4 (2017): 455–464; C. Y. Lo, V. Looi, W. F. Thompson, and C. M. McMahon, "Music Training for Children With Sensorineural Hearing Loss Improves Speech-in-Noise Perception," *Journal of Speech, Language, and Hearing Research* 63, no. 6 (2020): 1990–2015.

Chapter 6

1. N. L. Wallin, B. Merker, and S. Brown, *The Origins of Music* (Cambridge, MA: MIT Press, 2000).

2. A. B. Lord, *The Singer of Tales* (Cambridge, MA: Harvard University Press, 1960).

3. T. Gioia, *Work Songs* (Durham, NC: Duke University Press, 2006).

4. H. Pham, "West Africa Ghana, Post Office," YouTube, June 22, 2011, https://www.youtube.com/watch?v=c3fctmixsKE.

5. M. Aminian, *The Woven Sounds* (documentary film). 2019.

6. S. Brown and J. Jordania, "Universals in the World's Musics," *Psychology of Music* 41, no. 2 (2011): 229–248.

7. S. Dehaene, *Consciousness and the Brain: Deciphering How the Brain Codes Our Thoughts* (New York: Viking, 2014).

8. S. A. Kotz, A. Ravignani, and W. T. Fitch, "The Evolution of Rhythm Processing," *Trends in Cognitive Science* 22, no. 10 (2018): 896–910.

9. A. Tierney and N. Kraus, "Neural Entrainment to the Rhythmic Structure of Music," *Journal of Cognitive Neuroscience* 27, no. 2 (2015): 400–408.

10. I. J. Moon, S. Kang, N. Boichenko, S. H. Hong, and K. M. Lee, "Meter Enhances the Subcortical Processing of Speech Sounds at a Strong Beat," *Scientific Reports* 10, no. 1 (2020): 15973.

11. W. Fries and A. A. Swihart, "Disturbance of Rhythm Sense Following Right Hemisphere Damage," *Neuropsychologia* 28, no. 12 (1990): 1317–1323; M. Di Pietro, M. Laganaro, B. Leemann, and A. Schnider, "Receptive Amusia: Temporal Auditory Processing Deficit in a Professional Musician Following a Left Temporo-Parietal Lesion," *Neuropsychologia* 42, no. 7 (2004): 868–877; I. Peretz, "Processing of Local and Global Musical Information by Unilateral Brain-Damaged Patients," *Brain* 113, no. 4 (1990): 1185–1205; C. Liégeois-Chauvel, I. Peretz, M. Babai, V. Laguitton, and P. Chauvel, "Contribution of Different Cortical Areas in the Temporal Lobes to Music Processing," *Brain* 121, no. 10) (1998): 1853–1867.

12. A. Tierney and N. Kraus, "Evidence for Multiple Rhythmic Skills," *PLoS One* 10, no. 9 (2015): e0136645; S. Bonacina, J. Krizman, T. White-Schwoch, T. Nicol, and N. Kraus, "How Rhythmic Skills Relate and Develop in School-Age Children," *Global Pediatric Health* 6 (2019): 2333794X19852045.

13. A. Tierney, T. White-Schwoch, J. MacLean, and N. Kraus, "Individual Differences in Rhythm Skills: Links with Neural Consistency and Linguistic Ability," *Journal of Cognitive Neuroscience* 29, no. 5 (2017): 855–868; J. M. Thomson and U. Goswami, "Rhythmic Processing in Children with Developmental Dyslexia: Auditory and Motor

Rhythms Link to Reading and Spelling," *Journal of Physiology* 102, no. 1–3 (2008): 120–129; S. Bonacina, J. Krizman, T. White-Schwoch, and N. Kraus, "Clapping in Time Parallels Literacy and Calls Upon Overlapping Neural Mechanisms in Early Readers," *Annals of the New York Academy of Sciences* 1423 (2018): 338–348.

14. J. Slater, N. Kraus, K. W. Carr, A. Tierney, A. Azem, and R. Ashley, "Speech-in-Noise Perception Is Linked to Rhythm Production Skills in Adult Percussionists and Non-Musicians," *Language, Cognition and Neuroscience* 33, no. 6 (2018): 710–717.

15. A. Tierney and N. Kraus, "Getting Back on the Beat: Links between Auditory-Motor Integration and Precise Auditory Processing at Fast Time Scales," *European Journal of Neuroscience* 43, no. 6 (2016): 782–791.

16. A. A. Benasich, Z. Gou, N. Choudhury, and K. D. Harris, "Early Cognitive and Language Skills Are Linked to Resting Frontal Gamma Power across the First 3 Years," *Behavioural Brain Research* 195, no. 2 (2008): 215–222.

17. J. M. Thomson and U. Goswami, "Rhythmic Processing in Children with Developmental Dyslexia: Auditory and Motor Rhythms Link to Reading and Spelling," *Journal of Physiology* 102, no. 1–3 (2008): 120–129; P. Wolff, "Timing Precision and Rhythm in Developmental Dyslexia," *Reading and Writing* 15 (2002): 179–206; J. Thomson, B. Fryer, J. Maltby, and U. Goswami, "Auditory and Motor Rhythm Awareness in Adults with Dyslexia," *Journal of Research in Reading* 29 (2006): 334–348; K. H. Corriveau and U. Goswami, "Rhythmic Motor Entrainment in Children with Speech and Language Impairments: Tapping to the Beat," *Cortex* 45, no. 1 (2009): 119–130; A. T. Tierney and N. Kraus, "The Ability to Tap to a Beat Relates to Cognitive, Linguistic, and Perceptual Skills," *Brain and Language* 124, no. 3 (2013): 225–231; C. S. Moritz, S. Yampolsky, G. Papadelis, J. Thomson, and M. Wolf, "Links between Early Rhythm Skills, Musical Training, and Phonological Awareness," *Reading and Writing* 26 (2013): 739–769.

18. P. Wolff, "Timing Precision and Rhythm in Developmental Dyslexia," *Reading and Writing* 15 (2002): 179–206.

19. A. T. Tierney and N. Kraus, "The Ability to Tap to a Beat Relates to Cognitive, Linguistic, and Perceptual Skills," *Brain and Language* 124, no. 3 (2013): 225–231.

20. K. Woodruff Carr, T. White-Schwoch, A. T. Tierney, D. L. Strait, and N. Kraus, "Beat Synchronization Predicts Neural Speech Encoding and Reading Readiness in Preschoolers," *Proceedings of the National Academy of Sciences of the United States of America* 111, no. 40 (2014): 14559–14564; S. Bonacina, J. Krizman, T. White-Schwoch, T. Nicol, and N. Kraus, "Distinct Rhythmic Abilities Align with Phonological Awareness and Rapid Naming in School-age Children," *Cognitive Processing* 21 (2020): 575–581; S. Bonacina, J. Krizman, T. White-Schwoch, and N. Kraus, "Clapping in Time Parallels Literacy and Calls upon Overlapping Neural Mechanisms in Early Readers." Annals of the New York Academy of Sciences 1423 (2018): 338–348.

21. K. J. Kohler, "Rhythm in Speech and Language: A New Research Paradigm," *Phonetica* 66, no. 1–2 (2009): 29–45.

22. J. Slater, N. Kraus, K. W. Carr, A. Tierney, A. Azem, and R. Ashley, "Speech-in-Noise Perception Is Linked to Rhythm Production Skills in Adult Percussionists and Non-Musicians," *Language, Cognition and Neuroscience* 33, no. 6 (2018): 710–717.

23. N. Kraus and T. White-Schwoch, "Neurobiology of Everyday Communication: What Have We Learned from Music?" *Neuroscientist* 23, no. 3 (2017): 287–298; A. Parbery-Clark, E. Skoe, C. Lam, and N. Kraus, "Musician Enhancement for Speech-in-Noise," *Ear and Hearing* 30, no. 6 (2009): 653–661; A. Parbery-Clark, D. L. Strait, S. Anderson, E. Hittner, and N. Kraus, "Musical Experience and the Aging Auditory System: Implications for Cognitive Abilities and Hearing Speech in Noise," *PLoS One* 6, no. 5 (2011): e18082; A. Parbery-Clark, A. Tierney, D. Strait, and N. Kraus, "Musicians Have Fine-Tuned Neural Distinction of Speech Syllables," *Neuroscience* 219 (2012): 111–119; B. R. Zendel and C. Alain, "Musicians Experience Less Age-Related Decline in Central Auditory Processing," *Psychology and Aging* 27, no. 2 (2012): 410–417; D. L. Strait, A. Parbery-Clark, E. Hittner, and N. Kraus, "Musical Training During Early Childhood Enhances the Neural Encoding of Speech in Noise," *Brain and Language* 123, no. 3 (2012): 191–201; J. Swaminathan, C. R. Mason, T. M. Streeter, V. Best, G. Kidd Jr., and A. D. Patel, "Musical Training, Individual Differences and the Cocktail Party Problem," *Scientific Reports* 5 (2015): 11628; B. R. Zendel, C. D. Tremblay, S. Belleville, and I. Peretz, "The Impact of Musicianship on the Cortical Mechanisms Related to Separating Speech from Background Noise," *Journal of Cognitive Neuroscience* 27, no. 5 (2015): 1044–1059.

24. A. D. Patel, J. R. Iversen, M. R. Bregman, and I. Schulz, "Experimental Evidence for Synchronization to a Musical Beat in a Nonhuman Animal," *Current Biology* 19, no. 10 (2009): 827–830.

25. S. M. Wilson, A. P. Saygin, M. I. Sereno, and M. Iacoboni, "Listening to Speech Activates Motor Areas Involved in Speech Production," *Nature Neuroscience* 7, no. 7 (2004): 701–702; S. C. Herholz, E. B. Coffey, C. Pantev, and R. J. Zatorre, "Dissociation of Neural Networks for Predisposition and for Training-Related Plasticity in Auditory-Motor Learning," *Cerebral Cortex* 26, no. 7 (2016): 3125–3134.

26. M. Bangert, T. Peschel, G. Schlaug, M. Rotte, D. Drescher, H. Hinrichs, H. J. Heinze, and E. Altenmuller, "Shared Networks for Auditory and Motor Processing in Professional Pianists: Evidence from Fmri Conjunction," *NeuroImage* 30, no. 3 (2006): 917–926.

27. M. Larsson, S. R. Ekstrom, and P. Ranjbar, "Effects of Sounds of Locomotion on Speech Perception," *Noise and Health* 17, no. 77 (2015): 227–232.

28. I. Winkler, G. P. Haden, O. Ladinig, I. Sziller, and H. Honing, "Newborn Infants Detect the Beat in Music," *Proceedings of the National Academy of Sciences of the United States of America* 106, no. 7 (2009): 2468–2471.

29. J. Phillips-Silver and L. J. Trainor, "Feeling the Beat: Movement Influences Infant Rhythm Perception," *Science* 308, no. 5727 (2005): 1430.

30. M. J. Hove and J. L. Risen, "It's All in the Timing: Interpersonal Synchrony Increases Affiliation," *Social Cognition* 27, no. 6 (2009): 949–961.

31. S. Kirschner and M. Tomasello, "Joint Drumming: Social Context Facilitates Synchronization in Preschool Children," *Journal of Experimental Child Psychology* 102, no. 3 (2009): 299–314.

32. L. K. Cirelli, K. M. Einarson, and L. J. Trainor, "Interpersonal Synchrony Increases Prosocial Behavior in Infants," *Developmental Science* 17, no. 6 (2014): 1003–1011.

33. Y. Hou, B. Song, Y. Hu, Y. Pan, and Y. Hu, "The Averaged Inter-Brain Coherence between the Audience and a Violinist Predicts the Popularity of Violin Performance," *NeuroImage* 211 (2020): 116655.

34. Musicians Without Borders, www.musicianswithoutborders.org.

35. T. Gioia, *Healing Songs* (Durham, NC: Duke University Press, 2006).

36. G. Reynolds, "Phys Ed: Does Music Make You Exercise Harder?" *New York Times*, August 25, 2010.

37. H. A. Lim, "Effect of 'Developmental Speech and Language Training through Music' on Speech Production in Children with Autism Spectrum Disorders," *Journal of Music Therapy* 47, no. 1 (2010): 2–26.

38. L. A. Nelson, M. Macdonald, C. Stall, and R. Pazdan, "Effects of Interactive Metronome Therapy on Cognitive Functioning After Blast-Related Brain Injury: A Randomized Controlled Pilot Trial," *Neuropsychology* 27, no. 6 (2013): 666–679; S. Hegde, "Music-Based Cognitive Remediation Therapy for Patients with Traumatic Brain Injury," *Frontiers in Neurology* 5 (2014): 34; M. H. Thaut, J. C. Gardiner, D. Holmberg, J. Horwitz, L. Kent, G. Andrews, B. Donelan, and G. R. McIntosh, "Neurologic Music Therapy Improves Executive Function and Emotional Adjustment in Traumatic Brain Injury Rehabilitation," *Annals of the New York Academy of Sciences* 1169 (2009): 406–416.

39. C. Nombela, L. E. Hughes, A. M. Owen, and J. A. Grahn, "Into the Groove: Can Rhythm Influence Parkinson's Disease?" *Neuroscience and Biobehavioral Reviews* 37, no. 10 Pt. 2 (2013): 2564–2570; M. J. de Dreu, A. S. van der Wilk, E. Poppe, G. Kwakkel, and E. E. van Wegen, "Rehabilitation, Exercise Therapy and Music in Patients with Parkinson's Disease: A Meta-Analysis of the Effects of Music-Based Movement Therapy on Walking Ability, Balance and Quality of Life," *Parkinsonism & Related Disorders* 18, Suppl. 1 (2012): S114–119; J. M. Hausdorff, J. Lowenthal, T. Herman, L. Gruendlinger, C. Peretz, and N. Giladi, "Rhythmic Auditory Stimulation Modulates

Gait Variability in Parkinson's Disease," *European Journal of Neuroscience* 26, no. 8 (2007): 2369–2375.

40. C. M. Tomaino, "Recovery of Fluent Speech Through a Musician's Use of Pre-learned Song Repertoire: A Case Study," *Music and Medicine* 2, no. 2 (2010): 85–88; C. M. Tomaino, "Effective Music Therapy Techniques in the Treatment of Nonfluent Aphasia," *Annals of the New York Academy of Sciences* 1252, no. 1 (2012): 312–317; E. L. Stegemoller, T. R. Hurt, M. C. O'Connor, R. D. Camp, C. W. Green, J. C. Pattee, and E. K. Williams, "Experiences of Persons with Parkinson's Disease Engaged in Group Therapeutic Singing," *Journal of Music Therapy* 54, no. 4 (2018): 405–431; A. Raglio, O. Oasi, M. Gianotti, A. Rossi, K. Goulene, and M. Stramba-Badiale, "Improvement of Spontaneous Language in Stroke Patients with Chronic Aphasia Treated with Music Therapy: A Randomized Controlled Trial," *International Journal of Neuroscience* 126, no. 3 (2016): 235–242; M. H. Thaut and G. C. McIntosh, "Neurologic Music Therapy in Stroke Rehabilitation," *Current Physical Medicine and Rehabilitation Reports* 2, no. 2 (2014): 106–113; C. M. Tomaino, "Clinical Applications of Music Therapy in Neurologic Rehabilitation," in *Music That Works*, R. B. Haas, pp. 211–20 (Austria: Springer-Verlag, 2009); J. P. Brady, "Metronome-Conditioned Speech Retraining for Stuttering," *Behavior Therapy* 2, no. 2 (1971): 129–150.

41. M. W. Hardy and A. B. Lagasse, "Rhythm, Movement, and Autism: Using Rhythmic Rehabilitation Research as a Model for Autism," *Frontiers in Integrative Neuroscience* 7 (2013): 19; A. B. Lagasse, "Effects of a Music Therapy Group Intervention on Enhancing Social Skills in Children with Autism," *Journal of Music Therapy* 51, no. 3 (2014): 250–275; A. B. Lagasse, "Social Outcomes in Children with Autism Spectrum Disorder: A Review of Music Therapy Outcomes," *Patient Related Outcome Measures* 8 (2017): 23–32.

42. L. K. Cirelli, K. M. Einarson, and L. J. Trainor, "Interpersonal Synchrony Increases Prosocial Behavior in Infants," *Developmental Science* 17, no. 6 (2014): 1003–1011.

43. S. Bonacina, J. Krizman, T. White-Schwoch, and N. Kraus, "Clapping in Time Parallels Literacy and Calls Upon Overlapping Neural Mechanisms in Early Readers," *Annals of the New York Academy of Sciences* 1423 (2018): 338–348; M. Ritter, K. A. Colson, and J. Park, "Reading Intervention Using Interactive Metronome in Children with Language and Reading Impairment: A Preliminary Investigation," *Communication Disorders Quarterly* 34, no. 2 (2012): 106–119; G. E. Taub, K. S. McGrew, and T. Z. Keith, "Improvements in Interval Time Tracking and Effects on Reading Achievement," *Psychology in the Schools* 44, no. 8 (2007): 849–963.

44. F. S. Barrett, H. Robbins, D. Smooke, J. L. Brown, and R. R. Griffiths, "Qualitative and Quantitative Features of Music Reported to Support Peak Mystical Experiences During Psychedelic Therapy Sessions," *Frontiers in Psychology* 8 (2017): 1238.

45. T. Gioia, *Healing Songs* (Durham, NC: Duke University Press, 2006).

46. W. R Thompson, S. S. Yen, and J. Rubin, "Vibration Therapy: Clinical Applications in Bone," *Current Opinion in Endocrinology, Diabetes, and Obesity* 21, no. 6 (2014): 447–453.

47. E. Muggenthaler, "The Felid Purr: A Healing Mechanism?" *Journal of the Acoustical Society of America* 110 (2001): 2666.

Chapter 7

1. E. Paulesu, E. McCrory, F. Fazio, L. Menoncello, N. Brunswick, S. F. Cappa, M. Cotelli, et al., "A Cultural Effect on Brain Function," *Nature Neuroscience* 3, no. 1 (2000): 91–96.

2. P. H. Seymour, M. Aro, and J. M. Erskine, "Foundation Literacy Acquisition in European Orthographies," *British Journal of Psychology* 94, part 2 (2003): 143–174; N. C. Ellis, M. Natsume, K. Stavropoulou, L. Hoxhallari, V. H. P. Daal, N. Polyzoe, M.-L. Tsipa, and M. Petalas, "The Effects of Orthographic Depth On Learning to Read Alphabetic, Syllabic, and Logographic Scripts," *Reading Research Quarterly* 39, no. 4 (2004): 438–468.

3. J. C. Ziegler, C. Perry, A. Ma-Wyatt, D. Ladner, and G. Schulte-Körne, "Developmental Dyslexia in Different Languages: Language-Specific or Universal?" *Journal of Experimental Child Psychology* 86, no. 3 (2003): 169–193; E. Paulesu, J. F. Demonet, F. Fazio, E. McCrory, V. Chanoine, N. Brunswick, S. F. Cappa, et al., "Dyslexia: Cultural Diversity and Biological Unity," *Science* 291, no. 5511 (2001): 2165–2167.

4. M. Wolf and C. J. Stoodley, *Proust and the Squid: The Story and Science of the Reading Brain* (New York: HarperCollins, 2007).

5. J. Stein, "The Magnocellular Theory of Developmental Dyslexia," *Dyslexia* 7, no. 1 (2001): 12–36; S. Singleton and S. Trotter, "Visual Stress in Adults with and without Dyslexia," *Journal of Research in Reading* 28, no. 3 (2005): 365–378; J. Stein, "The Current Status of the Magnocellular Theory of Developmental Dyslexia," *Neuropsychologia* 130 (2019): 66–77; S. M. Handler and W. M. Fierson, "Learning Disabilities, Dyslexia, and Vision," *Pediatrics* 127, no. 3 (2011): e818–856; P. Harries, R. Hall, N. Ray, and J. Stein, "Using Coloured Filters to Reduce the Symptoms of Visual Stress in Children with Reading Delay," *Scandinavian Journal of Occupational Therapy* 22, no. 2 (2015): 153–160.

6. A. A. Benasich and R. H. Fitch, *Developmental Dyslexia: Early Precursors, Neurobehavioral Markers and Biological Substrates* (Baltimore: Paul H. Brookes, 2012).

7. T. Teinonen, V. Fellman, R. Näätänen, P. Alku, and M. Huotilainen, "Statistical Language Learning in Neonates Revealed by Event-Related Brain Potentials," *BMC Neuroscience* 10 (2009): 21.

8. T. Teinonen, V. Fellman, R. Näätänen, P. Alku, and M. Huotilainen, "Statistical Language Learning in Neonates Revealed by Event-Related Brain Potentials," *BMC Neuroscience* 10 (2009): 21; J. R. Saffran, R. N. Aslin, and E. L. Newport, "Statistical Learning by 8-Month-Old Infants," *Science* 274, no. 5294 (1996): 1926–1928.

9. E. Skoe and N. Kraus, "Hearing It Again and Again: On-Line Subcortical Plasticity in Humans," *PLoS One* 5, no. 10 (2010): e13645.

10. B. Chandrasekaran, J. Hornickel, E. Skoe, T. Nicol, and N. Kraus, "Context-Dependent Encoding in the Human Auditory Brainstem," *Neuron* 64 (2009): 311–319.

11. H. M. Sigurdardottir, H. B. Danielsdottir, M. Gudmundsdottir, K. H. Hjartarson, E. A. Thorarinsdottir, and A. Kristjansson, "Problems with Visual Statistical Learning in Developmental Dyslexia," *Scientific Reports* 7, no. 1 (2017): 606; J. L. Evans, J. R. Saffran, and K. Robe-Torres, "Statistical Learning in Children with Specific Language Impairment," *Journal of Speech, Language, and Hearing Research* 52, no. 2 (2009): 321–335.

12. C. M. Conway, D. B. Pisoni, E. M. Anaya, J. Karpicke, and S. C. Henning, "Implicit Sequence Learning in Deaf Children with Cochlear Implants," *Developmental Science* 14, no. 1 (2011): 69–82.

13. A. A. Scott-Van Zeeland, K. McNealy, A. T. Wang, M. Sigman, S. Y. Bookheimer, and M. Dapretto, "No Neural Evidence of Statistical Learning During Exposure to Artificial Languages in Children with Autism Spectrum Disorders," *Biological Psychiatry* 68, no. 4 (2010): 345–351.

14. K. McNealy, J. C. Mazziotta, and M. Dapretto, "Age and Experience Shape Developmental Changes in the Neural Basis of Language-Related Learning," *Developmental Science* 14, no. 6 (2011): 1261–1282; J. Bartolotti, V. Marian, S. R. Schroeder, and A. Shook, "Bilingualism and Inhibitory Control Influence Statistical Learning of Novel Word Forms," *Frontiers in Psychology* 2 (2011): 324; A. Shook, V. Marian, J. Bartolotti, and S. R. Schroeder, "Musical Experience Influences Statistical Learning of a Novel Language," *American Journal of Psychology* 126, no. 1 (2013): 95–104; P. Vasuki R. M., M. Sharma, R. Ibrahim, and J. Arciuli, "Statistical Learning and Auditory Processing in Children with Music Training: An ERP Study," *Clinical Neurophysiology* 128, no. 7 (2017): 1270–1281; D. Schön and C. François, "Musical Expertise and Statistical Learning of Musical and Linguistic Structures," *Frontiers in Psychology* 2 (2011): 167.

15. L. Kishon-Rabin, O. Amir, Y. Vexler, and Y. Zaltz, "Pitch Discrimination: Are Professional Musicians Better Than Non-Musicians?" *Journal of Basic and Clinical Physiology and Pharmacology* 12, no. 2 (2001): 125–143; M. F. Spiegel and C. S. Watson, "Performance on Frequency-Discrimination Tasks by Musicians and Non-musicians," *Journal of the Acoustical Society of America* 76, no. 6 (1984): 1690–1695.

16. K. Banai and M. Ahissar, "Poor Frequency Discrimination Probes Dyslexics with Particularly Impaired Working Memory," *Audiology and Neurotology* 9, no. 6 (2004):

328–340; L. F. Halliday and D. V. Bishop, "Is Poor Frequency Modulation Detection Linked to Literacy Problems? A Comparison of Specific Reading Disability and Mild to Moderate Sensorineural Hearing Loss," *Brain and Language* 97, no. 2 (2006): 200–213; S. J. France, B. S. Rosner, P. C. Hansen, C. Calvin, J. B. Talcott, A. J. Richardson, and J. F. Stein, "Auditory Frequency Discrimination in Adult Developmental Dyslexics," *Perception and Psychophysics* 64, no. 2 (2002): 169–179.

17. P. Helenius, K. Uutela, and R. Hari, "Auditory Stream Segregation in Dyslexic Adults," *Brain* 122, part 5 (1999): 907–913.

18. J. B. Talcott, C. Witton, M. F. McLean, P. C. Hansen, A. Rees, G. G. Green, and J. F. Stein, "Dynamic Sensory Sensitivity and Children's Word Decoding Skills," *Proceedings of the National Academy of Sciences of the United States of America* 97, no. 6 (2000): 2952–2957.

19. T. Baldeweg, A. Richardson, S. Watkins, C. Foale, and J. Gruzelier, "Impaired Auditory Frequency Discrimination in Dyslexia Detected with Mismatch Evoked Potentials," *Annals of Neurology* 45, no. 4 (1999): 495–503.

20. M. van Ingelghem, A. van Wieringen, J. Wouters, E. Vandenbussche, P. Onghena, and P. Ghesquiere, "Psychophysical Evidence for a General Temporal Processing Deficit in Children with Dyslexia," *Neuroreport* 12, no. 16 (2001): 3603–3637; M. J. Hautus, G. J. Setchell, K. E. Waldie, and I. J. Kirk, "Age-Related Improvements in Auditory Temporal Resolution in Reading-Impaired Children," *Dyslexia* 9, no. 1 (2003): 37–45; M. Sharma, S. C. Purdy, P. Newall, K. Wheldall, R. Beaman, and H. Dillon, "Electrophysiological and Behavioral Evidence of Auditory Processing Deficits in Children with Reading Disorder," *Clinical Neurophysiology* 117, no. 5 (2006): 1130–1144.

21. S. Rosen and E. Manganari, "Is There a Relationship between Speech and Nonspeech Auditory Processing in Children with Dyslexia?" *Journal of Speech, Language, and Hearing Research* 44, no. 4 (2001): 720–736.

22. P. Menell, K. I. McAnally, and J. F. Stein, "Psychophysical Sensitivity and Physiological Response to Amplitude Modulation in Adult Dyslexic Listeners," *Journal of Speech, Language, and Hearing Research* 42, no. 4 (1999): 797–803.

23. B. Boets, M. Vandermosten, H. Poelmans, H. Luts, J. Wouters, and P. Ghesquiere, "Preschool Impairments in Auditory Processing and Speech Perception Uniquely Predict Future Reading Problems," *Research in Developmental Disabililties* 32, no. 2 (2011): 560–570; K. H. Corriveau, U. Goswami, and J. M. Thomson, "Auditory Processing and Early Literacy Skills in a Preschool and Kindergarten Population," *Journal of Learning Disabilities* 43, no. 4 (2010): 369–382.

24. A. A. Benasich and P. Tallal, "Infant Discrimination of Rapid Auditory Cues Predicts Later Language Impairment," *Behavioural Brain Research* 136, no. 1 (2002): 31–49.

25. M. M. Merzenich, W. M. Jenkins, P. Johnston, C. Schreiner, S. L. Miller, and P. Tallal, "Temporal Processing Deficits of Language-Learning Impaired Children Ameliorated by Training," *Science* 271, no. 5245 (1996): 77–81; P. Tallal, S. L. Miller, G. Bedi, X. Wang, S. S. Nagarajan, C. Schreiner, W. M. Jenkins, and M. M. Merzenich, "Language Comprehension in Language-Learning Impaired Children Improved with Acoustically Modified Speech," *Science* 271, No. 5245 (1996): 81–84.

26. E. Temple, G. K. Deutsch, R. A. Poldrack, S. L. Miller, P. Tallal, M. M. Merzenich, and J. D. E. Gabrieli, "Neural Deficits in Children with Dyslexia Ameliorated by Behavioral Remediation: Evidence from Functional MRI," *Proceedings of the National Academy of Sciences of the United States of America* 100, no. 5 (2003): 2860–2855.

27. A. A. Benasich, N. A. Choudhury, T. Realpe-Bonilla, and C. P. Roesler, "Plasticity in Developing Brain: Active Auditory Exposure Impacts Prelinguistic Acoustic Mapping," *Journal of Neuroscience* 34, no. 40 (2014): 13349–13363.

28. P. Lieberman, R. H. Meskill, M. Chatillon, and H. Schupack, "Phonetic Speech Perception Deficits in Dyslexia," *Journal of Speech and Hearing Research* 28, no. 4 (1985): 480–486.

29. N. Kraus, T. J. McGee, T. D. Carrell, S. G. Zecker, T. G. Nicol, and D. B. Koch, "Auditory Neurophysiologic Responses and Discrimination Deficits in Children with Learning Problems," *Science* 273, no. 5277 (1996): 971–973.

30. P. Lieberman, R. H. Meskill, M. Chatillon, and H. Schupack, "Phonetic Speech Perception Deficits in Dyslexia," *Journal of Speech and Hearing Research* 28, no. 4 (1985): 480–486.

31. N. Kraus, T. J. McGee, T. D. Carrell, S. G. Zecker, T. G. Nicol, and D. B. Koch, "Auditory Neurophysiologic Responses and Discrimination Deficits in Children with Learning Problems," *Science* 273, no. 5277 (1996): 971–973.

32. C. King, C. M. Warrier, E. Hayes, and N. Kraus, "Deficits in Auditory Brainstem Encoding of Speech Sounds in Children with Learning Problems," *Neuroscience Letters* 319, no. (2002): 111–115; J. Cunningham, T. Nicol, S. G. Zecker, A. Bradlow, and N. Kraus, "Neurobiologic Responses to Speech in Noise in Children with Learning Problems: Deficits and Strategies for Improvement," *Clinical Neurophysiology* 112 (2001): 758–767.

33. B. Wible, T. Nicol, and N. Kraus, "Correlation between Brainstem and Cortical Auditory Processes in Normal and Language-Impaired Children," *Brain* 128 (2005): 417–423; B. Wible, T. Nicol, and N. Kraus, "Atypical Brainstem Representation of Onset and Formant Structure of Speech Sounds in Children with Language-Based Learning Problems," *Biological Psychology* 67 (2004): 299–317.

34. K. Banai, J. M. Hornickel, E. Skoe, T. Nicol, S. Zecker, and N. Kraus, "Reading and Subcortical Auditory Function," *Cerebral Cortex* 19, no. 11 (2009): 2699–2707.

35. E. Skoe, T. Nicol, and N. Kraus, "Cross-Phaseogram: Objective Neural Index of Speech Sound Differentiation," *Journal of Neuroscience Methods* 196, no. 2 (2011): 308–317; T. White-Schwoch and N. Kraus, "Physiologic Discrimination of Stop Consonants Relates to Phonological Skills in Pre-Readers: a Biomarker For Subsequent Reading Ability?" *Frontiers in Human Neuroscience* 7 (2013): 899.

36. G. A. Miller and P. E. Nicely, "An Analysis of Perceptual Confusions Among Some English Consonants," *Journal of the Acoustical Society of America* 27, no. 2 (1955): 338–52; J. Meyer, L. Dentel, and F. Meunier, "Speech Recognition in Natural Background Noise," *PLOS ONE* 8, no. 11 (2013): e79279.

37. J. Hornickel and N. Kraus, "Unstable Representation of Sound: A Biological Marker of Dyslexia," *Journal of Neuroscience* 33, no. 8 (2013): 3500–3504.

38. T. White-Schwoch, K. Woodruff Carr, E. C. Thompson, S. Anderson, T. Nicol, A. R. Bradlow, S. G. Zecker, and N. Kraus, "Auditory Processing in Noise: A Preschool Biomarker For Literacy," *PLoS Biology* 13, no. 7 (2015): e1002196.

39. The statistical modeling necessary to titrate the three ingredients into a powerful predictor was performed by Brainvolts' project-wide senior data analyst, Travis White-Schwoch, lead author on the paper that reported the finding.

40. T. White-Schwoch, K. Woodruff Carr, E. C. Thompson, S. Anderson, T. Nicol, A. R. Bradlow, S. G. Zecker, and N. Kraus, "Auditory Processing in Noise: A Preschool Biomarker for Literacy," *PLoS Biology* 13, no. 7 (2015): e1002196.

41. J. Hornickel, S. Zecker, A. Bradlow, and N. Kraus, "Assistive Listening Devices Drive Neuroplasticity in Children with Dyslexia," *Proceedings of the National Academy of Sciences of the United States of America* 109, no. 41 (2012): 16731–1636.

42. B. Hart and T. R. Risley, *Meaningful Differences in the Everyday Experience of Young American Children* (Baltimore: P. H. Brookes, 1995).

43. J. Gilkerson, J. A. Richards, S. F. Warren, J. K. Montgomery, C. R. Greenwood, D. Kimbrough Oller, J. H. L. Hansen, and T. D. Paul, "Mapping the Early Language Environment Using All-Day Recordings and Automated Analysis," *American Journal of Speech-Language Pathology* 26, no. 2 (2017): 248–265; D. E. Sperry, L. L. Sperry, and P. J. Miller, "Reexamining the Verbal Environments of Children from Different Socioeconomic Backgrounds," *Child Development* 90, no. 4 (2019): 1303–1318.

44. E. Hoff, "The Specificity of Environmental Influence: Socioeconomic Status Affects Early Vocabulary Development Via Maternal Speech," *Child Development* 74, no. 5 (2003): 1368–1378; E. Hoff-Ginsberg, "The Relation of Birth Order and Socioeconomic Status to Children's Language Experience and Language Development," *Applied Psycholinguistics* 19, no. 4 (1998): 603–629; J. Huttenlocher, H. Waterfall, M. Vasilyeva, J. Vevea, and L. V. Hedges, "Sources of Variability in Children's Language Growth," *Cognitive Psychology* 61, no. 4 (2010): 343–365; M. L. Rowe, "Child-Directed Speech: Relation

to Socioeconomic Status, Knowledge of Child Development and Child Vocabulary Skill," *Journal of Child Language* 35, no. 1 (2008): 185–205; A. Fernald, V. A. Marchman, and A. Weisleder, "SES Differences in Language Processing Skill and Vocabulary Are Evident At 18 Months," *Developmental Science* 16, no. 2 (2013): 234–248.

45. A. J. Tomarken, G. S. Dichter, J. Garber, and C. Simien, "Resting Frontal Brain Activity: Linkages to Maternal Depression and Socio-Economic Status Among Adolescents," *Biological Psychology* 67, no. 1–2 (2004): 77–102; R. D. Raizada, T. L. Richards, A. Meltzoff, and P. K. Kuhl, "Socioeconomic Status Predicts Hemispheric Specialisation of the Left Inferior Frontal Gyrus in Young Children," *NeuroImage* 40, no. 3 (2008): 1392–401; M. A. Sheridan, K. Sarsour, D. Jutte, M. D'Esposito, and W. T. Boyce, "The Impact of Social Disparity on Prefrontal Function in Childhood," *PLoS One* 7, no. 4 (2012): e35744.

46. K. G. Noble, S. M. Houston, E. Kan, and E. R. Sowell, "Neural Correlates of Socioeconomic Status in the Developing Human Brain," *Developmental Science* 15, no. 4 (2012): 516–527; J. L. Hanson, A. Chandra, B. L. Wolfe, and S. D. Pollak, "Association between Income and the Hippocampus," *PLoS One* 6, no. 5 (2011): e18712; K. Jednoróg, I. Altarelli, K. Monzalvo, J. Fluss, J. Dubois, C. Billard, G. Dehaene-Lambertz, and F. Ramus, "The Influence of Socioeconomic Status on Children's Brain Structure," *PLOS ONE* 7, no. 8 (2012): e42486.

47. J. Gilkerson, J. A. Richards, S. F. Warren, J. K. Montgomery, C. R. Greenwood, D. Kimbrough Oller, J. H. L. Hansen, and T. D. Paul, "Mapping the Early Language Environment Using All-Day Recordings and Automated Analysis," *American Journal of Speech-Language Pathology* 26, no. 2 (2017): 248–265; E. A. Cartmill, B. F. Armstrong III, L. R. Gleitman, S. Goldin-Meadow, T. N. Medina, and J. C. Trueswell, "Quality of Early Parent Input Predicts Child Vocabulary 3 Years Later," *Proceedings of the National Academy of Sciences of the United States of America* 110, no. 28 (2013): https://doi.org/10.1073/pnas.1309518110.

48. J. Huttenlocher, H. Waterfall, M. Vasilyeva, J. Vevea, and L. V. Hedges, "Sources of Variability in Children's Language Growth," *Cognitive Psychology* 61, no. 4 (2010): 343–365; M. L. Rowe, "A Longitudinal Investigation of the Role of Quantity and Quality of Child-Directed Speech in Vocabulary Development," *Child Development* 83, no. 5 (2012): 1762–1774; J. F. Schwab, and C. Lew-Williams, "Language Learning, Socioeconomic Status, and Child-Directed Speech," *Wiley Interdisciplinary Reviews: Cognitive Science* 7, no. 4 (2016): 264–275.

49. J. Gilkerson, and J. A. Richards. *The LENA Natural Language Study* (Boulder, CO: LENA Foundation, 2008).

50. K. Wong, C. Thomas, and M. Boben, "Providence Talks: A Citywide Partnership to Address Early Childhood Language Development," *Studies in Educational Evaluation* (2020): 64.

51. E. Skoe, J. Krizman, and N. Kraus, "The Impoverished Brain: Disparities in Maternal Education Affect the Neural Response to Sound," *Journal of Neuroscience* 33, no. 44 (2013): 17221–17231.

52. N. M. Russo, E. Skoe, B. Trommer, T. Nicol, S. Zecker, A. Bradlow, and N. Kraus, "Deficient Brainstem Encoding of Pitch in Children with Autism Spectrum Disorders," *Clinical Neurophysiology* 119, no. 8 (2008): 1720–1723.

53. D. A. Abrams, C. J. Lynch, K. M. Cheng, J. Phillips, K. Supekar, S. Ryali, L. Q. Uddin, and V. Menon, "Underconnectivity between Voice-Selective Cortex and Reward Circuitry in Children with Autism," *Proceedings of the National Academy of Sciences of the United States of America* 110, no. 29 (2013): 12060–12065.

54. C. Chevallier, G. Kohls, V. Troiani, E. S. Brodkin, and R. T. Schultz, "The Social Motivation Theory of Autism," *Trends in Cognitive Sciences* 16, no. 4 (2012): 231–239.

55. M. Font-Alaminos, M. Cornella, J. Costa-Faidella, A. Hervás, S. Leung, I. Rueda, and C. Escera, "Increased Subcortical Neural Responses to Repeating Auditory Stimulation in Children with Autism Spectrum Disorder," *Biological Psychology* (in press).

56. B. L. Maslen and J. R. Maslen, *Bob Books Series* (Scholastic: New York, 1976–).

57. W. I. Serniclaes, S. Van Heghe, P. Mousty, R. Carr, and L. Sprenger-Charolles, "Allophonic Mode of Speech Perception in Dyslexia," *Journal of Experimental Child Psychology* 87, no. 4 (2004): 336–361.

58. D. A. Treffert, "The Savant Syndrome: An Extraordinary Condition. A Synopsis: Past, Present, Future," *Philosophical Transactions of the Royal Society of London. Series B, Biological Sciences* 364, no. 1522 (2009): 1351–1357.

59. E. L. Grigorenko, A. Klin, D. L. Pauls, R. Senft, C. Hooper, and F. Volkmar, "A Descriptive Study of Hyperlexia in a Clinically Referred Sample of Children with Developmental Delays," *Journal of Autism and Developmental Disorders* 32, no. 1 (2002): 3–12.

60. J. M. Quinn and R. K. Wagner, "Gender Differences in Reading Impairment and in the Identification of Impaired Readers: Results from a Large-Scale Study of At-Risk Readers," *Journal of Learning Disabilities* 48, no. 4 (2015): 433–445; K. A. Flannery, J. Liederman, L. Daly, and J. Schultz, "Male Prevalence for Reading Disability Is Found in a Large Sample of Black and White Children Free from Ascertainment Bias," *Journal of the International Neuropsychological Society* 6, no. 4 (2000): 433–442.

61. J. I Benichov, S. E. Benezra, D. Vallentin, E. Globerson, M. A. Long, and O. Tchernichovski, "The Forebrain Song System Mediates Predictive Call Timing in Female and Male Zebra Finches," *Current Biology* 26, no. 3 (2016): 309–318.

62. C. Del Negro and J. M. Edeline, "Differences in Auditory and Physiological Properties of HVc Neurons between Reproductively Active Male and Female Canaries (*Serinus Canaria*)," *European Journal of Neuroscience* 14, no. 8 (2001): 1377–1389;

M. D. Gall, T. S. Salameh, and J. R. Lucas, "Songbird Frequency Selectivity and Temporal Resolution Vary with Sex and Season," *Proceedings of the Royal Society B: Biological Sciences* 280, no. 1751 (2013): 20122296.

63. J. A. Miranda, K. N. Shepard, S. K. McClintock, and R. C. Liu, "Adult Plasticity in the Subcortical Auditory Pathway of the Maternal Mouse," *PLoS One* 9, no. 7 (2014): e101630.

64. J. Krizman, S. Bonacina, and N. Kraus, "Sex Differences in Subcortical Auditory Processing Emerge Across Development," *Hearing Research* 380 (2019): 166–174.

65. J. Jerger and J. Hall, "Effects of Age and Sex on Auditory Brainstem Response," *Archives of Otolaryngology—Head and Neck Surgery* 106, no. 7 (1980): 387–391.

66. J. L. Krizman, S. Bonacina, N. Kraus "Sex Differences in Subcortical Auditory Processing Only Partially Explain Higher Prevalence of Language Disorders in Males," *Hearing Research* 398 (2020): 108075.

67. W. Kintsch and E. Kozminsky, "Summarizing Stories After Reading and Listening," *Journal of Educational Psychology* 69, no. 5 (1977): 491–499; B. A. Rogowsky, B. M. Calhoun, and P. Tallal, "Does Modality Matter? The Effects of Reading, Listening, and Dual Modality on Comprehension," *Sage Open* 6, no. 3 (2016); F. Deniz, A. O. Nunez-Elizalde, A. G. Huth, and J. L. Gallant, "The Representation of Semantic Information Across Human Cerebral Cortex During Listening Versus Reading Is Invariant to Stimulus Modality," *Journal of Neuroscience* 39, no. 39 (2019): 7722–7736.

68. C. M. MacLeod, K. Gopie, K. L. Hourihan, K. R. Neary, and J. D. Ozubko, "The Production Effect: Delineation of a Phenomenon," *Journal of Experimental Psychology: Learning, Memory, and Cognition* 36 (2010): 671–685; V. E. Pritchard, M. Heron-Delaney, S. A. Malone, and C. M. MacLeod, "The Production Effect Improves Memory in 7- to 10-Year-Old Children." *Child Development* 91, no. 3 (2020): 901–913.

Chapter 8

1. A. Parbery-Clark, E. Skoe, C. Lam, and N. Kraus, "Musician Enhancement for Speech-in-Noise," *Ear and Hearing* 30, no. 6 (2009): 653–661; B. R. Zendel and C. Alain, "Concurrent Sound Segregation Is Enhanced in Musicians," *Journal of Cognitive Neuroscience* 21, no. 8 (2009): 1488–1498; B. R. Zendel and C. Alain, "Musicians Experience Less Age-Related Decline in Central Auditory Processing," *Psychology and Aging* 27, no. 2 (2012): 410–417; G. M. Bidelman and A. Krishnan, "Effects of Reverberation on Brainstem Representation of Speech in Musicians and Non-Musicians," *Brain Research* 1355 (2010): 112–125; A. Parbery-Clark, E. Skoe, and N. Kraus, *Biological Bases for the Musician Advantage for Speech-in-Noise. Society for Neuroscience, Auditory Satellite* (Chicago: APAN, 2009); A. Parbery-Clark, E. Skoe, and N. Kraus, "Musical Experience Limits the Degradative Effects of Background Noise on

the Neural Processing of Sound," *Journal of Neuroscience* 29, no. 45 (2009): 14100–14107; A. Parbery-Clark, A. Tierney, D. Strait, and N. Kraus, "Musicians Have Fine-Tuned Neural Distinction of Speech Syllables," *Neuroscience* 219 (2012): 111–119; A. Tierney, J. Krizman, E. Skoe, K. Johnston, and N. Kraus, "High School Music Classes Enhance the Neural Processing of Speech," *Frontiers in Psychology* 4 (2013): 855; D. L. Strait, A. Parbery-Clark, E. Hittner, and N. Kraus, "Musical Training During Early Childhood Enhances the Neural Encoding of Speech in Noise," *Brain and Language* 123, no. 3 (2012): 191–201; D. L. Strait, A. Parbery-Clark, S. O'Connell, and N. Kraus, "Biological Impact of Preschool Music Classes onProcessing Speech in Noise," *Developmental Cognitive Neuroscience* 6 (2013): 51–60.

2. A. D. Patel, "Why Would Musical Training Benefit the Neural Encoding of Speech? The OPERA Hypothesis," *Frontiers in Psychology* 2 (2011): 142.

3. M. Forgeard, G. Schlaug, A. Norton, C. Rosam, U. Iyengar, and E. Winner, "The Relation between Music and Phonological Processing in Normal-Reading Children and Children with Dyslexia," *Music Perception* 25, no. 4 (2008): 383–390.

4. J. Slater, A. Tierney, and N. Kraus, "At-Risk Elementary School Children with One Year of Classroom Music Instruction Are Better at Keeping a Beat," *PLoS One* 8, no. 10 (2013): e77250.

5. M. Forgeard, G. Schlaug, A. Norton, C. Rosam, U. Iyengar, and E. Winner, "The Relation between Music and Phonological Processing in Normal-Reading Children and Children with Dyslexia," *Music Perception* 25, no. 4 (2008): 383–390; S. H. Anvari, L. J. Trainor, J. Woodside, and B. A. Levy, "Relations Among Musical Skills, Phonological Processing, and Early Reading Ability in Preschool Children," *Journal of Experimental Child Psychology* 83, no. 2 (2002): 111–130; M. Huss, J. P. Verney, T. Fosker, N. Mead, and U. Goswami, "Music, Rhythm, Rise Time Perception and Developmental Dyslexia: Perception of Musical Meter Predicts Reading and Phonology," *Cortex* 47, no. 6 (2011): 674–689; R. F. McGivern, C. Berka, M. L. Languis, and S. Chapman, "Detection of Deficits in Temporal Pattern Discrimination Using the Seashore Rhythm Test in Young Children with Reading Impairments," *Journal of Learning Disabilities* 24, no. 1 (1991): 58–62; B. W. Atterbury, "A Comparison of Rhythm Pattern Perception and Perfor mance in Normal and Learning-Disabled Readers, Age 7 and 8," *Journal of Research in Music Education* 31, no. 4 (1983): 259–270; G. Dellatolas, L. Watier, M. T. Le Normand, T. Lubart, and C. Chevrie-Muller, "Rhythm Reproduction in Kindergarten, Reading Performance at Second Grade, and Developmental Dyslexia Theories," *Archives of Clinical Neuropsychology* 24, no. 6 (2009): 555–563; C. Moritz, S. Yampolsky, G. Papadelis, J. Thomson, and M. Wolf, "Links between Early Rhythm Skills, Musical Training, and Phonological Awareness," *Reading and Writing* 26 (2013): 739–769; J. Thomson, B. Fryer, J. Maltby, and U. Goswami, "Auditory and Motor Rhythm Awareness in Adults with Dyslexia," *Journal of Research in Reading* 29 (2006): 334–348; J. M. Thomson and U. Goswami,

"Rhythmic Processing in Children with Developmental Dyslexia: Auditory and Motor Rhythms Link to Reading and Spelling," *Journal of Physiology* 102, no. 1–3 (2008): 120–129; K. H. Corriveau and U. Goswami, "Rhythmic Motor Entrainment in Children with Speech and Language Impairments: Tapping to the Beat," *Cortex* 45, no. 1 (2009): 119–130; D. David, L. Wade-Woolley, J. R. Kirby, and K. Smithrim, "Rhythm and Reading Development in School-Age Children: A Longitudinal Study," *Journal of Research in Reading* 30, no. 2 (2007): 169–183; P. Wolff, "Timing Precision and Rhythm in Developmental Dyslexia," *Reading and Writing* 15 (2002): 179–120.

6. C. Moritz, S. Yampolsky, G. Papadelis, J. Thomson, and M. Wolf, "Links between Early Rhythm Skills, Musical Training, and Phonological Awareness," *Reading and Writing* 26 (2013): 739–769; E. Flaugnacco, L. Lopez, C. Terribili, M. Montico, S. Zoia, and D. Schon, "Music Training Increases Phonological Awareness and Reading Skills in Developmental Dyslexia: A Randomized Control Trial," *PLOS ONE* 10, no. 9 (2015): e0138715; K. Overy, "Dyslexia and Music: From Timing Deficits to Musical Intervention," in *The Neurosciences and Music*, ed. G. Avanzini, C. Faienza, L. Lopez, M. Majno, and D. Minciacchi, 497–505 (New York: The New York Academy of Sciences, 2003); H. Cogo-Moreira, C. R. Brandão de Ávila, G. B. Ploubidis, and J. de Jesus Maria, "Effectiveness of Music Education for the Improvement of Reading Skills and Academic Achievement in Young Poor Readers: A Pragmatic Cluster-Randomized, Controlled Clinical Trial," *PLOS ONE* 8, no. 3 (2013): e59984; F. H. Rauscher and S. C. Hinton, "Music Instruction and Its Diverse Extra-Musical Benefits," *Music Perception* 29, no. 2 (2011): 215–226; L. Herrera, O. Lorenzo, S. Defior, G. Fernandez-Smith, and E. Costa-Giomi, "Effects of Phonological and Musical Training on the Reading Readiness of Native- and Foreign-Spanish-Speaking Children," *Psychology of Music* 39, no. 1 (2010): 68–81; F. Degé and G. Schwarzer, "The Effect of a Music Program onPhonological Awareness in Preschoolers," *Frontiers in Psychology* 2 (2011): 124.

7. E. Flaugnacco, L. Lopez, C. Terribili, M. Montico, S. Zoia, and D. Schon, "Music Training Increases Phonological Awareness and Reading Skills in Developmental Dyslexia: A Randomized Control Trial," *PLOS ONE* 10, no. 9 (2015): e0138715; H. Cogo-Moreira, C. R. Brandão de Ávila, G. B. Ploubidis, and J. de Jesus Maria, "Effectiveness of Music Education for the Improvement of Reading Skills and Academic Achievement in Young Poor Readers: A Pragmatic Cluster-Randomized, Controlled Clinical Trial," *PLOS ONE* 8, no. 3 (2013): e59984; D. Fisher, "Early Language Learning with and Without Music," *Reading Horizons* 42, no. 1 (2001); I. Hurwitz, P. H. Wolff, B. D. Bortnick, and K. Kokas, "Nonmusical Effects of Kodaly Music Curriculum in Primary Grade Children," *Journal of Learning Disabilities* 8, no. 3 (1975): 167–74; S. Douglas and P. Willatts, "The Relationship between Musical Ability and Literacy Skills," *Journal of Research in Reading* 17, no. 2 (1994): 99–107; M. Forgeard, E. Winner, A. Norton, and G. Schlaug, "Practicing a Musical Instrument in Childhood Is Associated with Enhanced Verbal Ability and Nonverbal Reasoning," *PLOS ONE* 3, no. 10 (2008):

e3566; S. Moreno, C. Marques, A. Santos, M. Santos, S. L. Castro, and M. Besson, "Musical Training Influences Linguistic Abilities in 8-Year-Old Children: More Evidence for Brain Plasticity," *Cerebral Cortex* 19, no. 3 (2009): 712–23; G. E. Taub and P. J. Lazarus, "The Effects of Training in Timing and Rhythm on Reading Achievment," *Contemporary Issues in Education Research* 5, no. 4 (2013): 343–350; I. Rautenberg, "The Effects of Musical Training on the Decoding Skills of German-Speaking Primary School Children," *Journal of Research in Reading* 38, no. 1 (2015): 1–17.

8. A. Tierney and N. Kraus, "The Ability to Move to a Beat Is Linked to the Consistency of Neural Responses to Sound," *Journal of Neuroscience* 33, no. 38 (2013): 14981–14988; K. Woodruff Carr, A. Tierney, T. White-Schwoch, and N. Kraus, "Intertrial Auditory Neural Stability Supports Beat Synchronization in Preschoolers," *Developmental Cognitive Neuroscience* 17 (2016): 76–82; N. Kraus, J. Slater, E. Thompson, J. Hornickel, D. Strait, T. Nicol, and T. White-Schwoch, "Music Enrichment Programs Improve the Neural Encoding of Speech in At-Risk Children," *Journal of Neuroscience* 34, no. 36 (2014): 11913–11918.

9. A. Parbery-Clark, E. Skoe, C. Lam, and N. Kraus, "Musician Enhancement for Speech-in-Noise," *Ear and Hearing* 30, no. 6 (2009): 653–61; B. R. Zendel and C. Alain, "Concurrent Sound Segregation Is Enhanced in Musicians," *Journal of Cognitive Neuroscience* 21, no. 8 (2009): 1488–1498; B. R. Zendel and C. Alain, "Musicians Experience Less Age-Related Decline in Central Auditory Processing," *Psychology and Aging* 27, no. 2 (2012): 410–17; G. M. Bidelman and A. Krishnan, "Effects of Reverberation on Brainstem Representation of Speech in Musicians and Non-Musicians," *Brain Research* 1355 (2010): 112–125; A. Parbery-Clark, E. Skoe, and N. Kraus, "Biological Bases for the Musician Advantage for Speech-in-Noise," presentation at Society for Neuroscience, Auditory Satellite (APAN), Chicago, 2009; A. Parbery-Clark, E. Skoe, and N. Kraus, "Musical Experience Limits the Degradative Effects of Background Noise on the Neural Processing of Sound," *Journal of Neuroscience* 29, no. 45 (2009): 14100–14107; A. Parbery-Clark, A. Tierney, D. Strait, and N. Kraus, "Musicians Have Fine-Tuned Neural Distinction of Speech Syllables," *Neuroscience* 219 (2012): 111–119; A. Tierney, J. Krizman, E. Skoe, K. Johnston, and N. Kraus, "High School Music Classes Enhance the Neural Processing of Speech," *Frontiers in Psychology* 4 (2013): 855; D. L. Strait, A. Parbery-Clark, E. Hittner, and N. Kraus, "Musical Training During Early Childhood Enhances the Neural Encoding of Speech in Noise," *Brain and Language* 123, no. 3 (2012): 191–201; D. L. Strait, A. Parbery-Clark, S. O'Connell, and N. Kraus, "Biological Impact of Preschool Music Classes on Processing Speech in Noise," *Developmental Cognitive Neuroscience* 6 (2013): 51–60; A. Parbery-Clark, E. Skoe, and N. Kraus, "Musical Experience Improves Speech-in-Noise Perception: Behavioural and Neurophysiological Evidence," presentation at Society for Music Perception and Cognition, Indianapolis, IN, 2009.

10. J. Slater, E. Skoe, D. L. Strait, S. O'Connell, E. Thompson, and N. Kraus, "Music Training Improves Speech-in-Noise Perception: Longitudinal Evidence from a Community-Based Music Program," *Behavioural Brain Research* 291 (2015): 244–252.

11. Y. Du and R. J. Zatorre, "Musical Training Sharpens and Bonds Ears and Tongue to Hear Speech Better," *Proceedings of the National Academy of Sciences of the United States of America* 114, no. 51 (2017): 13579–13584.

12. A. Parbery-Clark, E. Skoe, and N. Kraus, "Musical Experience Limits the Degradative Effects of Background Noise on the Neural Processing of Sound," *Journal of Neuroscience* 29, no. 45 (2009): 14100–14107.

13. J. Slater, N. Kraus, K. W. Carr, A. Tierney, A. Azem, and R. Ashley, "Speech-in-Noise Perception Is Linked to Rhythm Production Skills in Adult Percussionists and Non-Musicians," *Language, Cognition and Neuroscience* 33, no. 6 (2018): 710–717.

14. A. Parbery-Clark, E. Skoe, C. Lam, and N. Kraus, "Musician Enhancement for Speech-in-Noise," *Ear and Hearing* 30, no. 6 (2009): 653–661.

15. B. R. Zendel and C. Alain, "Concurrent Sound Segregation Is Enhanced in Musicians," *Journal of Cognitive Neuroscience* 21, no. 8 (2009): 1488–1498; B. R. Zendel and C. Alain, "Musicians Experience Less Age-Related Decline in Central Auditory Processing," *Psychology and Aging* 27, no. 2 (2012): 410–417; D. L. Strait, A. Parbery-Clark, E. Hittner, and N. Kraus, "Musical Training During Early Childhood Enhances the Neural Encoding of Speech in Noise," *Brain and Language* 123, no. 3 (2012): 191–201; A. Parbery-Clark, D. L. Strait, S. Anderson, E. Hittner, and N. Kraus, "Musical Experience and the Aging Auditory System: Implications for Cognitive Abilities and Hearing Speech in Noise," *PLoS One* 6, no. 5 (2011): e18082; B. Hanna-Pladdy and A. Mackay, "The Relation between Instrumental Musical Activity and Cognitive Aging," *Neuropsychology* 25, no. 3 (2011): 378–386.

16. P. C. M. Wong, E. Skoe, N. M. Russo, T. Dees, and N. Kraus, "Musical Experience Shapes Human Brainstem Encoding of Linguistic Pitch Patterns," *Nature Neuroscience* 10, no. 4 (2007): 420–422.

17. A. Parbery-Clark, D. L. Strait, and N. Kraus, "Context-Dependent Encoding in the Auditory Brainstem Subserves Enhanced Speech-in-Noise Perception in Musicians," *Neuropsychologia* 49, no. 12 (2011): 3338–3345; C. Francois and D. Schön, "Musical Expertise Boosts Implicit Learning of Both Musical and Linguistic Structures," *Cerebral Cortex* 21, no. 10 (2011): 2357–2365.

18. D. R. Ruggles, R. L. Freyman, and A. J. Oxenham, "Influence of Musical Training on Understanding Voiced and Whispered Speech in Noise," *PLOS ONE* 9, no. 1 (2014): e86980; D. Boebinger, S. Evans, S. Rosen, C. F. Lima, T. Manly, and S. K. Scott, "Musicians and Non-Musicians Are Equally Adept at Perceiving Masked Speech," *Journal of the Acoustical Society of America* 137, no. 1 (2015): 378–387.

19. E. Skoe and N. Kraus, "A Little Goes a Long Way: How the Adult Brain Is Shaped by Musical Training in Childhood," *Journal of Neuroscience* 32, no. 34 (2012): 11507–11510.

20. T. White-Schwoch, K. W. Carr, S. Anderson, D. L. Strait, and N. Kraus, "Older Adults Benefit from Music Training Early in Life: Biological Evidence for Long-Term Training-Driven Plasticity," *Journal of Neuroscience* 33, no. 45 (2013): 17667–17674; B. Hanna-Pladdy and A. Mackay, "The Relation between Instrumental Musical Activity and Cognitive Aging," *Neuropsychology* 23, no. 3 (2011): 378–386; M. A. Balbag, N. L. Pedersen and M. Gatz, "Playing a Musical Instrument as a Protective Factor against Dementia and Cognitive Impairment: A Population-Based Twin Study," *Internation Journal of Alzheimer's Disease* 2014 (2014): 836748; T. Amer, B. Kalender, L. Hasher, S. E. Trehub and Y. Wong, "Do Older Professional Musicians Have Cognitive Advantages?" *PLOS ONE* 8, no. 8 (2013): e71630.

21. A. Tierney, J. Krizman, E. Skoe, K. Johnston, and N. Kraus, "High School Music Classes Enhance the Neural Processing of Speech," *Frontiers in Psychology* 4 (2013): 855; A. T. Tierney, J. Krizman, and N. Kraus, "Music Training Alters the Course of Adolescent Auditory Development," *Proceedings of the National Academy of Sciences of the United States of America* 112, no. 32 (2015): 10062–10067.

22. J. Hornickel, E. Skoe, T. Nicol, S. Zecker, and N. Kraus, "Subcortical Differentiation of Stop Consonants Relates to Reading and Speech-in-Noise Perception," *Proceedings of the National Academy of Sciences of the United States of America* 106, no. 31 (2009): 13022–13027.

23. J. Chobert, C. François, J. L. Velay, and M. Besson, "Twelve Months of Active Musical Training in 8- to 10-Year-Old Children Enhances the Preattentive Processing of Syllabic Duration and Voice Onset Time," *Cerebral Cortex* 24, no. 4 (2014): 956–967.

24. S. Moreno, C. Marques, A. Santos, M. Santos, S. L. Castro, and M. Besson, "Musical Training Influences Linguistic Abilities in 8-Year-Old Children: More Evidence for Brain Plasticity," *Cerebral Cortex* 19, no. 3 (2009): 712–723; S. Moreno and M. Besson, "Influence of Musical Training on Pitch Processing: Event-Related Brain Potential Studies of Adults and Children," *Annals of the New York Academy of Sciences* 1060 (2005): 93–97; S. Moreno, E. Bialystok, R. Barac, E. G. Schellenberg, N. J. Cepeda, and T. Chau, "Short-Term Music Training Enhances Verbal Intelligence and Executive Function," *Psychological Science* 22, no. 11 (2011): 1425–1433.

25. A. C. Jaschke, H. Honing, and E. J. A. Scherder, "Longitudinal Analysis of Music Education on Executive Functions in Primary School Children," *Frontiers in Neuroscience* 12 (2018): 103.

26. A. Habibi, B. R. Cahn, A. Damasio, and H. Damasio, "Neural Correlates of Accelerated Auditory Processing in Children Engaged in Music Training," *Developmental Cognitive Neuroscience* 21 (2016): 1–14.

27. H. Yang, W. Ma, D. Gong, J. Hu, and D. Yao, "A Longitudinal Study on Children's Music Training Experience and Academic Development," *Scientific Reports* 4 (2014): 5854.

28. T. Linnavalli, V. Putkinen, J. Lipsanen, M. Huotilainen, and M. Tervaniemi, "Music Playschool Enhances Children's Linguistic Skills," *Scientific Reports* 8, no. 1 (2018): 8767.

29. M. L. Whitson, S. Robinson, K. V. Valkenburg, and M. Jackson, "The Benefits of an Afterschool Music Program for Low-Income, Urban Youth: the Music Haven Evaluation Project," *Journal of Community Psychology* (forthcoming).

30. S. L. Hennessy, M. E. Sachs, B. Ilari, and A. Habibi, "Effects of Music Training onInhibitory Control and Associated Neural Networks in School-Aged Children: a Longitudinal Study," *Frontiers in Neuroscience* 13 (2019): 1080.

31. V. Putkinen, M. Tervaniemi, K. Saarikivi, P. Ojala, and M. Huotilainen, "Enhanced Development of Auditory Change Detection in Musically Trained School-Aged Children: a Longitudinal Event-Related Potential Study. *Developmental Science* 17, no. 2 (2014): 282–297; A. T. Tierney, J. Krizman, and N. Kraus, "Music Training Alters the Course of Adolescent Auditory Development," *Proceedings of the National Academy of Sciences of the United States of America* 112, no. 32 (2015): 10062–10067; A. Habibi, A. Damasio, B. Ilari, R. Veiga, A. Joshi, R. Leahy, J. Haldar, D. Varadarajan, C. Bhushan, and H. Damasio, "Childhood Music Training Induces Change in Micro and Macroscopic Brain Structure; Results from a Longitudinal Study," *Cerebral Cortex* 28, no. 12 (2018): 4336–4347; A. Habibi, R. B. Cahn, A. Damasio, and H. Damasio, "Neural Correlates of Accelerated Auditory Processing in Children Engaged in Music Training," *Developmental Cognitive Neuroscience* 21 (2016): 1–14; B. S. Ilari, P. Keller, H. Damasio, and A. Habibi, "The Development of Musical Skills of Underprivileged Children Over the Course of 1 Year: A Study in the Context of an El Sistema-Inspired Program," *Frontiers in Psychology* 7 (2016): 62.

32. A. J. Tomarken, G. S. Dichter, J. Garber, and C. Simien, "Resting Frontal Brain Activity: Linkages to Maternal Depression and Socio-Economic Status Among Adolescents. *Biological Psychology* 67, no. 1–2 (2004): 77–102; R. D. Raizada, T. L. Richards, A. Meltzoff, and P. K. Kuhl, "Socioeconomic Status Predicts Hemispheric Specialisation of the Left Inferior Frontal Gyrus in Young Children," *NeuroImage* 40, no. 3 (2008): 1392–1401; M. A. Sheridan, K. Sarsour, D. Jutte, M. D'Esposito, and W. T. Boyce, "The Impact of Social Disparity on Prefrontal Function in Child-hood," *PLoS One* 7, no. 4 (2012): e35744; K. G. Noble, S. M. Houston, E. Kan, and E. R. Sowell, "Neural Correlates of Socioeconomic Status in the Developing Human Brain," *Developmental Science* 15, no. 4 (2012): 516–527; J. L. Hanson, A. Chandra, B. L. Wolfe, and S. D. Pollak, "Association between Income and the Hippocampus," *PLoS One* 6, no. 5 (2011): e18712; K. Jednoróg, I. Altarelli, K. Monzalvo, J. Fluss, J. Dubois, C. Billard, G. Dehaene-Lambertz, and F. Ramus, "The Influence of Socioeco-nomic Status onChildren's Brain Structure," *PLoS One* 7, no. 8 (2012): e4248.

33. E. Skoe, J. Krizman, and N. Kraus, "The Impoverished Brain: Disparities in Maternal Education Affect the Neural Response to Sound," *Journal of Neuroscience* 33, no. 44 (2013): 17221–1731.

34. M. Lacour and L. D. Tissington, "The Effects of Poverty on Academic Achievement," *Educational Research Review* 7, no. 6 (2011): 522–527.

35. K. E. Stanovich, "Matthew Effects in Reading—Some Consequences of Individual-Differences in the Acquisition of Literacy," *Reading Research Quarterly* 21, no. 4 (1986): 360–407.

36. J. Slater, D. Strait, E. Skoe, S. O'Connell, E. Thompson, and N. Kraus, "Longitudinal Effects of Group Music Instruction onLiteracy Skills in Low Income Children," *PLOS ONE* 9, no. 11 (2014): e113383.

37. S. Saarikallio and J. Erkkilä, "The Role of Music in Adolescents' Mood Regulation," *Psychology of Music* 35 (2007): 88–109; S. Saarikallio, "Music as Emotional Self-Regulation Throughout Adulthood," *Psychology of Music* 39, no. 3 (2011): 307–327.

38. N. Mammarella, B. Fairfield, and C. Cornoldi, "Does Music Enhance Cognitive Perfor mance in Healthy Older Adults? the Vivaldi Effect" *Aging Clinical and Experimental Research* 19, no. 5 (2007): 394–399; H. C. Beh and R. Hirst, "Performance on Driving-Related Tasks During Music," *Ergonomics* 42, no. 8 (1999): 1087–1098; S. Hallam, J. Price, and G. Katsarou, "The Effects of Background Music on Primary School Pupils' Task Performance," *Educational Studies* 28, no. 2 (2002): 111–122.

39. L. Ferreri, E. Mas-Herrero, R. J. Zatorre, P. Ripolles, A. Gomez-Andres, H. Alicart, G. Olive, et al., "Dopamine Modulates the Reward Experiences Elicited by Music," *Proceedings of the National Academy of Sciences of the United States of America* 116, no. 9 (2019): 3793–3798.

40. F. G. Ashby, A. M. Isen, and U. Turken, "A Neuropsychological Theory of Positive Affect and Its Influence on Cognition," *Psychological Review* 106, no. 3 (1999): 529–550.

41. T. Särkämö and D. Soto, "Music Listening After Stroke: Beneficial Effects and Potential Neural Mechanisms," *Annals of the New York Academy of Sciences* 1252 (2012): 266–281.

42. N. Kraus, J. Slater, E. Thompson, J. Hornickel, D. Strait, T. Nicol, and T. White-Schwoch, "Auditory Learning Through Active Engagement with Sound: Biological Impact of Community Music Lessons in At-Risk Children," *Frontiers in Neuroscience* 8 (2014): 351.

43. N. Kraus, J. Slater, E. Thompson, J. Hornickel, D. Strait, T. Nicol, and T. White-Schwoch, "Music Enrichment Programs Improve the Neural Encoding of Speech in At-Risk Children," *Journal of Neuroscience* 34, no. 36 (2014): 11913–18; J. Slater, E. Skoe, D. L. Strait, S. O'Connell, E. Thompson, and N. Kraus, "Music Training Improves Speech-in-Noise Perception: Longitudinal Evidence from a Community-Based Music Program," *Behavioural Brain Research* 291 (2015): 244–252.

44. M. L. Fermanich, "Money for Music Education: A District Analysis of the How, What, and Where of Spending for Music Education," *Journal of Education Finance* 37, no. 2 (2011): 130–149.

45. N. Kraus and T. White-Schwoch, "The Argument for Music Education," *American Scientist* 108 (2020): 210–213.

46. J. Daugherty, "Why Music Matters: The Cognitive Personalism of Reimer and Elliott," *Australian Journal of Music Education* 1 (1996): 29–37.

47. B. Reimer, *A Philosophy of Music Education* (Englewood Cliffs, NJ: Prentice-Hall, 1970).

48. A. D. Patel, "Evolutionary Music Cognition: Cross-species Studies," in *Foundations in Music Psychology: Theory and Research*, ed. P. J. Rentfrow and D. Levitin (Cambridge, MA: MIT Press, 2019), 459–501.

49. I. Peretz, *How Music Sculpts Our Brain* (Paris/New York: Odile Jacob, 2019).

50. D. Elliott, *Music Matters: A Philosophy of Music Education* (New York: Oxford University Press, 1995).

51. J. Slater, A. Azem, T. Nicol, B. Swedenborg, and N. Kraus, "Variations on the Theme of Musical Expertise: Cognitive and Sensory Processing in Percussionists, Vocalists and Non-Musicians," *European Journal of Neuroscience* 45, no. 7 (2017): 952–956.

52. V. Mongelli, S. Dehaene, F. Vinckier, I. Peretz, P. Bartolomeo, and L. Cohen, "Music and Words in the Visual Cortex: The Impact of Musical Expertise," *Cortex* 86 (2017): 260–274; F. Bouhali, V. Mongelli, M. Thiebaut de Schotten, and L. Cohen, "Reading Music and Words: The Anatomical Connectivity of Musicians' Visual Cortex," *NeuroImage* 212 (2020): 116666.

Chapter 9

1. F. Grosjean, "Individual Bilingualism," in *The Encyclopedia of Language and Linguistics*, ed. R. E. Asher and J. M. Y. Simpson (Oxford: Pergamon Press, 1994).

2. R. Näätänen, A. Lehtokoski, M. Lennes, M. Cheour, M. Huotilainen, A. Iivonen, M. Vainio, P. Alku, R. J. Ilmoniemi, A. Luuk, J. Allik, J. Sinkkonen, and K. Alho, "Language-Specific Phoneme Representations Revealed by Electric and Magnetic Brain Responses," *Nature* 385, no. 6615 (1997): 432–434.

3. C. Ryan, *Language Use in the United States: 2011* (Washington, DC: US Census Bureau, 2013).

4. D. J. Saer, "The Effect of Bilingualism on Intelligence," *British Journal of Psychology* 14, no. 1 (1923): 25–38.

5. G. G. Thompson, *Child Psychology; Growth Trends in Psychological Adjustment* (Boston: Houghton Mifflin, 1952).

6. K. Hakuta, *Mirror of Language: The Debate on Bilingualism* (New York: Basic Books, 1986).

7. A. Sharma and M. F. Dorman, "Neurophysiologic Correlates of Cross-Language Phonetic Perception," *Journal of the Acoustical Society of America* 107, no. 5, part 1 (2000): 2697–2703.

8. A. Sharma and M. F. Dorman, "Neurophysiologic Correlates of Cross-Language Phonetic Perception," *Journal of the Acoustical Society of America* 107, no. 5, part 1 (2000): 2697–2703.

9. A. M. Liberman, K. S. Harris, H. S. Hoffman, and B. C. Griffith, "The Discrimination of Speech Sounds Within and Across Phoneme Boundaries," *Journal of Experimental Psychology* 54, no. 5 (1957): 358–368.

10. K. Tremblay, N. Kraus, T. J. McGee, C. W. Ponton, and B. Otis, "Central Auditory Plasticity: Changes in the N1-P2 Complex After Speech-Sound Training," *Ear and Hearing* 22, no. 2 (2001): 79–90; A. R. Bradlow, D. B. Pisoni, R. Akahane-Yamada, and Y. Tohkura, "Training Japanese Listeners to Identify English /R/ and /L/: IV. Some Effects of Perceptual Learning on Speech Production," *Journal of the Acoustical Society of America* 101, no. 4 (1997): 2299–2310.

11. A. R. Bradlow, R. Akahane-Yamada, D. B. Pisoni, and Y. Tohkura, "Training Japanese Listeners to Identify English /R/ and /L/: Long-Term Retention of Learning in Perception and Production," *Perception and Psychophysics* 61, no. 5 (1999): 977–985.

12. R. Näätänen, A. Lehtokoski, M. Lennes, M. Cheour, M. Huotilainen, A. Iivonen, M. Vainio, P. Alku, R. J. Ilmoniemi, A. Luuk, J. Allik, J. Sinkkonen, and K. Alho, "Language-Specific Phoneme Representations Revealed by Electric and Magnetic Brain Responses," *Nature* 385, no. 6615 (1997): 432–434.

13. B. Chandrasekaran, A. Krishnan, and J. T. Gandour, "Mismatch Negativity to Pitch Contours Is Influenced by Language Experience," *Brain Research* 1128, no. 1 (2007): 148–156.

14. M. Cheour, R. Ceponiene, A. Lehtokoski, A. Luuk, J. Allik, K. Alho, and R. Näätänen, "Development of Language-Specific Phoneme Representations in the Infant Brain," *Nature Neuroscience* 1, no. 5 (1998): 351–353.

15. P. K. Kuhl, S. Kiritani, T. Deguchi, A. Hayashi, E. B. Stevens, C. D. Dugger, and P. Iverson, "Effects of Language Experience on Speech Perception: American and Japanese Infants' Perception of /Ra/ and /La/," *Journal of the Acoustical Society of America* 102, no. 5 (1997): 3135.

16. P. K. Kuhl, K. A. Williams, F. Lacerda, K. N. Stevens, and B. Lindblom, "Linguistic Experience Alters Phonetic Perception in Infants by 6 Months of Age," *Science* 255, no. 5044 (1992): 606–608.

17. C. M. Weber-Fox and H. J. Neville, "Maturational Constraints on Functional Specializations For Language Processing: ERP and Behavioral Evidence in Bilingual Speakers," *Journal of Cognitive Neuroscience* 8, no. 3 (1996): 231–56; V. Marian, M. Spivey, and J. Hirsch, "Shared and Separate Systems in Bilingual Language Processing: Converging Evidence from Eyetracking and Brain Imaging," *Brain and Language* 86, no. 1 (2003): 70–82; H. Sumiya and A. F. Healy, "Phonology in the Bilingual Stroop Effect," *Memory and Cognition* 32, no. 5 (2004): 752–758.

18. A. Rodriguez-Fornells, A. van der Lugt, M. Rotte, B. Britti, H. J. Heinze, and T. F. Munte, "Second Language Interferes with Word Production in Fluent Bilinguals: Brain Potential and Functional Imaging Evidence," *Journal of Cognitive Neuroscience* 17, no. 3 (2005): 422–433.

19. M. J. Spivey and V. Marian, "Cross Talk between Native and Second Languages: Partial Activation of an Irrelevant Lexicon," *Psychological Science* 10, no. 3 (1999): 281–284.

20. G. Thierry and Y. J. Wu, "Brain Potentials Reveal Unconscious Translation During Foreign-Language Comprehension," *Proceedings of the National Academy of Sciences of the United States of America* 104, no. 30 (2007): 12530–12535.

21. E. Bialystok, *Bilingualism in Development: Language, Literacy, and Cognition* (Cambridge: Cambridge University Press, 2001).

22. P. M. Roberts, L. J. Garcia, A. Desrochers, and D. Hernandez, "English Performance of Proficient Bilingual Adults on the Boston Naming Test," *Aphasiology* 16, no. 4–6 (2002): 635–645; J. S. Portocarrero, R. G. Burright, and P. J. Donovick, "Vocabulary and Verbal Fluency of Bilingual and Monolingual College Students," *Archives of Clinical Neuropsychology* 22, no. 3 (2007): 415–422.

23. M. Kaushanskaya and V. Marian, "Bilingual Language Processing and Interference in Bilinguals: Evidence from Eye Tracking and Picture Naming," *Language Learning* 57, no. 1 (2007): 119–163; G. M. Bidelman and L. Dexter, "Bilinguals at the 'Cocktail Party': Dissociable Neural Activity in Auditory-Linguistic Brain Regions Reveals Neurobiological Basis for Nonnative Listeners' Speech-in-Noise Recognition Deficits," *Brain and Language* 143 (2015): 32–41; C. L. Rogers, J. J. Lister, D. M. Febo, J. M. Besing, and H. B. Abrams, "Effects of Bilingualism, Noise, and Reverberation on Speech Perception by Listeners with Normal Hearing," *Applied Psycholinguistics* 27, no. 3 (2006): 465–485; L. H. Mayo, M. Florentine, and S. Buus, "Age of Second-Language Acquisition and Perception of Speech in Noise," *Journal of Speech, Language, and Hearing Research* 40, no. 3 (1997): 686–693.

24. M. L. Garcia Lecumberri, M. Cooke, and A. Cutler, "Non-Native Speech Perception in Adverse Conditions: a Review," *Speech Communication* 52, no. 11–12 (2010): 864–886.

25. P. A. Luce and D. B. Pisoni, "Recognizing Spoken Words: the Neighborhood Activation Model," *Ear and Hearing* 19, no. 1 (1998): 1–36.

26. J. Krizman, A. R. Bradlow, S. S. Y. Lam, and N. Kraus, "How Bilinguals Listen in Noise: Linguistic and Non-Linguistic Factors," *Bilingualism: Language and Cognition* 20, no. 4 (2017): 834–843.

27. A. S. Dick, N. L. Garcia, S. M. Pruden, W. K. Thompson, S. W. Hawes, M. T. Sutherland, M. C. Riedel, A. R. Laird, and R. Gonzalez, "No Evidence for a Bilingual Executive Function Advantage in the Nationally Representative ABCD Study," *Nature Human Behavior* 3, no. 7 (2019): 692–701; K. R. Paap, H. A. Johnson, and O. Sawi, "Bilingual Advantages in Executive Functioning Either Do Not Exist or Are Restricted to Very Specific and Undetermined Circumstances," *Cortex* 69 (2015): 265–278.

28. E. Bialystok and M. M. Martin, "Attention and Inhibition in Bilingual Children: Evidence from the Dimensional Change Card Sort Task," *Developmental Science* 7, no. 3 (2014): 325–339; A. Costa, M. Hernández, and N. Sebastián-Gallés, "Bilingualism Aids Conflict Resolution: Evidence from the ANT Task" *Cognition* 106, no. 1 (2008): 59–86; E. Bialystok, "Cognitive Complexity and Attentional Control in the Bilingual Mind," *Child Development* 70, no. 3 (1999): 636–644; J. Krizman, V. Marian, A. Shook, E. Skoe, and N. Kraus, "Subcortical Encoding of Sound Is Enhanced in Bilinguals and Relates to Executive Function Advantages," *Proceedings of the National Academy of Sciences of the United States of America* 109, no. 20 (2012): 7877–7881.

29. E. Bialystok, "Cognitive Complexity and Attentional Control in the Bilingual Mind," *Child Development* 70, no. 3 (1999): 636–44; H. K. Blumenfeld and V. Marian, "Bilingualism Influences Inhibitory Control in Auditory Comprehension," *Cognition* 118, no. 2 (2011): 245–257; A. Hartanto and H. Yang, "Does Early Active Bilingualism Enhance Inhibitory Control and Monitoring? A Propensity-Matching Analysis," *Journal of Experimental Psychology: Learning, Memory, and Cognition* 45, no. 2 (2019): 360–378; S. M. Carlson and A. N. Meltzoff, "Bilingual Experience and Executive Functioning in Young Children," *Developmental Science* 11, no. 2 (2008): 282–298.

30. D. M. Antovich and K. Graf Estes, "Learning Across Languages: Bilingual Experience Supports Dual Language Statistical Word Segmentation," *Developmental Science* 21, no. 2 (2018).

31. T. Wang and J. R. Saffran, "Statistical Learning of a Tonal Language: The Influence of Bilingualism and Previous Linguistic Experience," *Frontiers in Psychology* 5 (2014): 953; J. Bartolotti, V. Marian, S. R. Schroeder, and A. Shook, "Bilingualism and Inhibitory Control Influence Statistical Learning of Novel Word Forms," *Frontiers in Psychology* 2 (2011): 324.

32. J. Bartolotti and V. Marian, "Bilinguals' Existing Languages Benefit Vocabulary Learning in a Third Language," *Language Learning* 67, no. 1 (2017): 110–140.

33. C. M. Conway, D. B. Pisoni, and W. G. Kronenberger, "The Importance of Sound for Cognitive Sequencing Abilities: The Auditory Scaffolding Hypothesis," *Current Directions in Psychological Science* 18, no. 5 (2009): 275–279.

34. M. A. Gremp, J. A. Deocampo, A. M. Walk, and C. M. Conway, "Visual Sequential Processing and Language Ability in Children Who Are Deaf or Hard of Hearing," *Journal of Child Language* 46, no. 4 (2019): 785–799; P. C. Hauser, J. Lukomski, and T. Hillman, "Development of Deaf and Hard-of-Hearing Students' Executive Function," in *Deaf Cognition: Foundations and Outcomes,* ed. M. Marschark and P. Hauser, 286–308 (New York: Oxford University Press, 2008); D. B. Pisoni and M. Cleary, "Learning, Memory, and Cognitive Processes in Deaf Children Following Cochlear Implantation," in *Cochlear Implants: Auditory Prostheses and Electric Hearing,* ed. F.-G. Zeng, A. N. Popper, and R. R. Fay, 377–426 (New York: Springer, 2004); L. S. Davidson, A. E. Geers, S. Hale, M. M. Sommers, C. Brenner, and B. Spehar, "Effects of Early Auditory Deprivation on Working Memory and Reasoning Abilities in Verbal and Visuospatial Domains for Pediatric Cochlear Implant Recipients," *Ear and Hearing* 40, no. 3 (2019): 517–528; S. V. Bharadwaj and J. A. Mehta, "An Exploratory Study of Visual Sequential Processing in Children with Cochlear Implants," *International Journal of Pediatric Otorhinolaryngology* 85 (2016): 158–165.

35. E. Bialystok, F. I. Craik, R. Klein, and M. Viswanathan, "Bilingualism, Aging, and Cognitive Control: Evidence from the Simon Task," *Psychology and Aging* 19, no. 2 (2004): 290–303.

36. J. Krizman, V. Marian, A. Shook, E. Skoe, and N. Kraus, "Subcortical Encoding of Sound Is Enhanced in Bilinguals and Relates to Executive Function Advantages," *Proceedings of the National Academy of Sciences of the United States of America* 109, no. 20 (2012): 7877–7881; J. Krizman, J. Slater, E. Skoe, V. Marian, and N. Kraus, "Neural Processing of Speech in Children Is Influenced by Extent of Bilingual Experience," *Neuroscience Letters* 585 (2015): 48–53.

37. J. Krizman, J. Slater, E. Skoe, V. Marian, and N. Kraus, "Neural Processing of Speech in Children Is Influenced by Extent of Bilingual Experience," *Neuroscience Letters* 585 (2015): 48–53; J. Krizman, E. Skoe, V. Marian, and N. Kraus, "Bilingualism Increases Neural Response Consistency and Attentional Control: Evidence for Sensory and Cognitive Coupling," *Brain and Language* 128, no. 1 (2014): 34–40.

38. T. D. Hanley, J. C. Snidecor, and R. L. Ringel, "Some Acoustic Differences Among Languages," *Phonetica* 14 (1966): 97–107.

39. B. Lee and D. V. L. Sidtis, "The Bilingual Voice: Vocal Characteristics When Speaking Two Languages Across Speech Tasks," *Speech, Language and Hearing* 20, no. 3 (2017): 174–185.

40. J. Krizman, E. Skoe, and N. Kraus, "Bilingual Enhancements Have No Socioeconomic Boundaries," *Developmental Science* 19, no. 6 (2016): 881–891.

41. S. M. Carlson and A. N. Meltzoff, "Bilingual Experience and Executive Functioning in Young Children," *Developmental Science* 11, no. 2 (2008): 282–298.

42. W. C. So, "Cross-Cultural Transfer in Gesture Frequency in Chinese-English Bilinguals," *Language and Cognitive Processes* 25, no. 10 (2010): 1335–1353.

43. G. Stam, "Thinking for Speaking About Motion: L1 and L2 Speech and Gesture," *International Journal of Applied Linguistics* 44, no. 2 (2006).

44. M. Gullberg, "Bilingualism and Gesture," in *The Handbook of Bilingualism and Multilingualism,* ed. T. K. Bhatia and W. C. Ritchie (Hoboken, NJ: Wiley-Blackwell, 2013), 417–437.

45. B. de Gelder and M. J. Huis In 'T Veld, "Cultural Differences in Emotional Expressions and Body Language," in *The Oxford Handbook of Cultural Neuroscience,* ed. J. Y. Chiao, R. Seligman, and R. Turner (Oxford: Oxford University Press, 2016).

46. C. L. Caldwell-Harris, "Emotionality Differences between a Native and Foreign Language: Theoretical Implications," *Frontiers in Psychology* 5 (2014): 1055.

47. M. H. Bond and T. M. Lai, "Embarrassment and Code-Switching into a Second Language," *Journal of Social Psychology* 126, no. 2 (1986): 179–186.

Chapter 10

1. M. Naguib and K. Riebel, "Singing in Space and Time: The Biology of Birdsong," in *Biocommunication of Animals,* ed. G. Witzany (Dordrecht: Springer Science+Business, 2014), 233–247.

2. S. Nowicki, D. Hasselquist, S. Bensch, and S. Peters, "Nestling Growth and Song Repertoire Size in Great Reed Warblers: Evidence for Song Learning as an Indicator Mechanism in Mate Choice," *Proceedings of the Royal Society B: Biological Sciences* 267, no. 1460 (2000): 2419–2424.

3. E. D. Jarvis, "Learned Birdsong and the Neurobiology of Human Language," *Annals of the New York Academy of Sciences* 1016 (2004): 749–777.

4. E. P. Kingsley, C. M. Eliason, T. Riede, Z. Li, T. W. Hiscock, M. Farnsworth, S. L. Thomson, F. Goller, C. J. Tabin, and J. A. Clarke, "Identity and Novelty in the Avian Syrinx," *Proceedings of the National Academy of Sciences of the United States of America* 115, no. 41 (2018): 10209–10217.

5. R. A. Suthers, E. Vallet, A. Tanvez, and M. Kreutzer, "Bilateral Song Production in Domestic Canaries," *Journal of Neurobiology* 60, no. 3 (2004): 381–393.

6. C. P. Elemans, I. L. Spierts, U. K. Muller, J. L. Van Leeuwen, and F. Goller, "Bird Song: Superfast Muscles Control Dove's Trill," *Nature* 431, no. 7005 (2004): 146.

7. W. A. Calder, "Respiration During Song in the Canary (Serinus Canaria)," *Comparative Biochemistry and Physiology* 32, no. 2 (1970): 251–258.

8. J. M. Wild, F. Goller, and R. A. Suthers, "Inspiratory Muscle Activity During Bird Song," *Journal of Neurobiology* 36, no. 3 (1998): 441–453.

9. E. A. Armstrong, *A Study of Bird Song* (London: Oxford University Press, 1963).

10. C. Safina, *Becoming Wild: How Animal Cultures Raise Families, Create Beauty, and Achieve Peace* (New York: Henry Holt, 2020).

11. R. E. Lemon, "How Birds Develop Song Dialects," *Condor* 77, no. 4 (1975): 385–406; P. Marler and M. Tamura, "Song 'Dialects' in Three Populations of White-Crowned Sparrows," *Condor* 64 (1962): 368–377.

12. M. C. Baker, K. J. Spitler-Nabors, and D. C. Bradley, "Early Experience Determines Song Dialect Responsiveness of Female Sparrows," *Science* 214, no. 4522 (1981): 819–821.

13. E. L. Doolittle, B. Gingras, D. M. Endres, and W. T. Fitch, "Overtone-Based Pitch Selection in Hermit Thrush Song: Unexpected Convergence with Scale Construction in Human Music," *Proceedings of the National Academy of Sciences of the United States of America* 111, no. 46 (2014): 16616–16621.

14. A. A. Saunders, "Octaves and Kilocycles in Bird Songs," *Wilson Bulletin* 71 (1959): 280–282.

15. A. H. Wing "Notes on the Song Series of a Hermit Thrush in the Yukon," *The Auk* 68, no. 2 (1951): 189–193; C. Hartshorne, *Born to Sing: An Interpretation and World Survey of Bird Song* (Bloomington: Indiana University Press, 1973).

16. E. L. Doolittle, B. Gingras, D. M. Endres, and W. T. Fitch, "Overtone-Based Pitch Selection in Hermit Thrush Song: Unexpected Convergence with Scale Construction in Human Music," *Proceedings of the National Academy of Sciences of the United States of America* 111, no. 46 (2014): 16616–16621.

17. M. Araya-Salas, "Is Birdsong Music?" *Significance* 9, no. 6 (2012): 4–7.

18. L. F. Baptista and R. A. Keister, "Why Birdsong Is Sometimes Like Music," *Perspectives in Biology and Medicine* 48, no. 3 (2005): 426–443.

19. L. F. Baptista and R. A. Keister, "Why Birdsong Is Sometimes Like Music," *Perspectives in Biology and Medicine* 48, no. 3 (2005): 426–443.

20. E. A. Armstrong, *A Study of Bird Song* (London: Oxford University Press, 1963).

21. A. T. Tierney, F. A. Russo, and A. D. Patel, "The Motor Origins of Human and Avian Song Structure," *Proceedings of the National Academy of Sciences of the United States of America* 108, no. 37 (2011): 15510–15515.

22. E. Doolittle, "Music Theory Is for the Birds," *Conrad Grebel Review* 33, no. 2 (2015): 238–248.

23. W. Young and V. Arlington, "Translating the Language of Birds," *Verbatim* 28, no. 1 (2003): 1–5.

24. Y. Chen, L. E. Matheson, and J. T. Sakata, "Mechanisms Underlying the Social Enhancement of Vocal Learning in Songbirds," *Proceedings of the National Academy of Sciences of the United States of America* 113, no. 24 (2016): 6641–6646.

25. P. Marler, "A Comparative Approach to Vocal Learning—Song Development in White-Crowned Sparrows," *Journal of Comparative and Physiological Psychology* 71, no. 2 (1970): 1.

26. W. H. Thorpe, "The Learning of Song Patterns by Birds, with Especial Reference to the Song of the Chaffinch Fringilla Coelebs," *Ibis* 100 (1958): 535–570.

27. R. Dooling and M. Searcy, "Early Perceptual Selectivity in the Swamp Sparrow," *Developmental Psychobiology* 13, no. 5 (1980): 499–506.

28. J. M. Moore and S. M. N. Woolley, "Emergent Tuning for Learned Vocalizations in Auditory Cortex," *Nature Neuroscience* 22, no. 9 (2019): 1469–1476.

29. D. A. Nelson and P. Marler, "Innate Recognition of Song in White-Crowned Sparrows—a Role in Selective Vocal Learning," *Animal Behaviour* 46, no. 4 (1993): 806–808.

30. R. F. Braaten and K. Reynolds, "Auditory Preference for Conspecific Song in Isolation-Reared Zebra Finches," *Animal Behaviour* 58, no. 1 (1999): 105–111.

31. H. Lee, "In Birds' Songs, Brains and Genes, He Finds Clues to Speech: Interview with Erich Jarvis," *Quanta Magazine, January 30, 2018.*

32. P. K. Kuhl, S. Kiritani, T. Deguchi, A. Hayashi, E. B. Stevens, C. D. Dugger, and P. Iverson, "Effects of Language Experience on Speech Perception: American and Japanese Infants' Perception of /Ra/ and /La/," *Journal of the Acoustical Society of America* 102, no. 5 (1997): 3135; P. K. Kuhl, K. A. Williams, F. Lacerda, K. N. Stevens, and B. Lindblom, "Linguistic Experience Alters Phonetic Perception in Infants by 6 Months of Age," *Science* 255, no. 5044 (1991): 606–608.

33. P. K. Kuhl, F. M. Tsao, and H. M. Liu, "Foreign-Language Experience in Infancy: Effects of Short-Term Exposure and Social Interaction on Phonetic Learning," *Proceedings of the National Academy of Sciences of the United States of America* 100, no. 15 (2003): 9096–9101.

34. S. Coren, "Do Dogs Have a Musical Sense?" *Psychology Today*, April 2, 2012, https://www.psychologytoday.com/us/blog/canine-corner/201204/do-dogs-have -musical-sense.

35. M. R. Bregman, A. D. Patel, and T. Q. Gentner, "Songbirds Use Spectral Shape, Not Pitch, for Sound Pattern Recognition," *Proceedings of the National Academy of Sciences* 113, no. 6 (2016): 1666–1671.

36. S. H. Hulse, A. H. Takeuchi, and R. F. Braaten, "Perceptual Invariances in the Comparative Psychology of Music," *Music Perception* 10, no. 2 (1992): 151–184.

37. A. Bannerjee, S. M. Phelps, and M. A. Long, "Singing Mice," *Current Biology* 29 (2019): R183–R199.

38. E. D. Jarvis, "Learned Birdsong and the Neurobiology of Human Language," *Annals of the New York Academy of Sciences* 1016 (2004): 749–777.

39. S. Yanagihara and Y. Yazaki-Sugiyama, "Auditory Experience-Dependent Cortical Circuit Shaping for Memory Formation in Bird Song Learning," *Nature Communications* 7 (2016): 11946.

40. R. Mooney, "Neural Mechanisms for Learned Birdsong," *Learning and Memory* 16, no. 11 (2009): 655–669.

41. M. S. Brainard and A. J. Doupe, "What Songbirds Teach Us About Learning," *Nature* 417, no. 6886 (2002): 351–358.

42. E. P. Derryberry, J. N. Phillips, G. E. Derryberry, M. J. Blum, and D. Luther, "Singing in a Silent Spring: Birds Respond to a Half-Century Soundscape Reversion during the COVID-19 Shutdown," *Science* 370, no. 6516 (2020): 575–579.

43. P. Marler and S. Peters, "Long-Term Storage of Learned Birdsongs Prior to Production," *Animal Behaviour* 30 (1982): 479–482.

44. R. Mooney, "Neural Mechanisms for Learned Birdsong," *Learning & Memory* 16, no. 11 (2009): 655–669.

45. H. J. Leppelsack, "Critical Periods in Bird Song Learning," *Acta Oto-Laryngologica. Supplementum* 429 (1986): 57–60.

46. I. McGilchrist, *The Master and His Emissary: The Divided Brain and the Making of the Western World* (New Haven: Yale University Press, 2009).

47. M. L. Phan and D. S. Vicario, "Hemispheric Differences in Processing of Vocalizations Depend on Early Experience," *Proceedings of the National Academy of Sciences USA* 107, no. 5 (2010): 2301–6; H. U. Voss, K. Tabelow, J. Polzehl, O. Tchernichovski, K. K. Maul, D. Salgado-Commissariat, D. Ballon, and S. A. Helekar, "Functional MRI of the Zebra Finch Brain During Song Stimulation Suggests a Lateralized

Response Topography," *Proceedings of the National Academy of Sciences of the United States of America* 104, no. 25 (2007): 10667–10672.

48. M. J. West and A. P. King, "Female Visual Displays Affect the Development of Male Song in the Cowbird," *Nature* 334, no. 6179 (1988): 244–246.

49. J. Krizman, S. Bonacina, and N. Kraus, "Sex Differences in Subcortical Auditory Processing Emerge Across Development," *Hearing Research* 380 (2019): 166–174.

50. C. J. Limb and A. R. Braun, "Neural Substrates of Spontaneous Musical Performance: An FMRI Study of Jazz Improvisation," *PLoS One* 3, no. 2 (2008): e1679.

51. P. Marler, S. Peters, G. F. Ball, A. M. Dufty Jr., and J. C. Wingfield, "The Role of Sex Steroids in the Acquisition and Production of Birdsong," *Nature* 336, no. 6201 (1988): 770–772.

52. G. Ritchison, "Variation in the Songs of Female Black-Headed Grosbeaks," *Wilson Bulletin* 97, no. 1 (1985): 47–56.

53. A. E. Illes and L. Yunes-Jimenez, "A Female Songbird Out-Sings Male Conspecifics During Simulated Territorial Intrusions," *Proceedings of the Royal Society B: Biological Sciences* 276, no. 1658 (2009): 981–986.

54. W. H. Webb, D. H. Brunton, J. D. Aguirre, D. B. Thomas, M. Valcu, and J. Dale, "Female Song Occurs in Songbirds with More Elaborate Female Coloration and Reduced Sexual Dichromatism," *Frontiers in Ecology and Evolution* 4 (2016): 22.

55. C. Safina, *Becoming Wild: How Animal Cultures Raise Families, Create Beauty, and Achieve Peace* (New York: Henry Holt, 2020).

Chapter 11

1. The National Institute for Occupational Safety and Health, "Occupational Noise Exposure: Revised Criteria, 1998," *U.S. Department of Health and Human Services* (1998): 98–126.

2. M. Chasin, *Hear the Music: Hearing Loss Prevention for Musicians*, 4th ed. (Toronto: Musicians' Clinics of Canada, 2010).

3. S. Cohen, G. W. Evans, D. S. Krantz, and D. Stokols, "Physiological, Motivational, and Cognitive Effects of Aircraft Noise on Children: Moving from the Laboratory to the Field," *American Psychologist* 35, no. 3 (1980): 231–243; M. M. Haines, S. A. Stansfeld, R. F. Job, B. Berglund, and J. Head, "Chronic Aircraft Noise Exposure, Stress Responses, Mental Health and Cognitive Performance in School Children," *Psychological Medicine* 31, no. 2 (2001): 265–277; S. A. Stansfeld, B. Berglund, C. Clark, I. Lopez-Barrio, P. Fischer, E. Ohrstrom, M. M. Haines, J. Head, S. Hygge, J. van Kamp, B. F. Berry, and RANCH Study Team, "Aircraft and Road Traffic Noise

and Children's Cognition and Health: A Cross-National Study," *Lancet* 365, no. 9475 (2005): 1942–1949; E. E. van Kempen, I. van Kamp, R. K. Stellato, I. Lopez-Barrio, M. M. Haines, M. E. Nilsson, C. Clark, D. Houthuijs, B. Brunekreef, B. Berglund, and S. A. Stansfeld, "Children's Annoyance Reactions to Aircraft and Road Traffic Noise," *Journal of the Acoustical Society of America* 125, no. 2 (2009): 895–904; G. W. Evans, S. Hygge, and M. Bullinger, "Chronic Noise and Psychological Stress," *Psychological Science* 6, no. 6 (1995): 333–338; B. Griefahn and M. Spreng, "Disturbed Sleep Patterns and Limitation of Noise," *Noise Health* 6, no. 22 (2004): 27–33; M. Spreng, "Possible Health Effects of Noise Induced Cortisol Increase," *Noise Health* 2, no. 7 (2000): 59–64.

4. M. Basner, W. Babisch, A. Davis, M. Brink, C. Clark, S. Janssen, and S. Stansfeld, "Auditory and Non-Auditory Effects of Noise on Health," *Lancet* 383, no. 9925 (2014): 1325–1332.

5. A. L. Bronzaft and D. P. McCarthy, "The Effect of Elevated Train Noise on Reading Ability," *Environment and Behavior* 7 (1975): 517–528.

6. A. L. Bronzaft, "The Effect of a Noise Abatement Program on Reading Ability," *Environmental Psychology* 1 (1981): 215–222.

7. M. P. Walker, *Why We Sleep: Unlocking the Power of Sleep and Dreams* (New York: Scribner, 2017).

8. M. Basner, W. Babisch, A. Davis, M. Brink, C. Clark, S. Janssen, and S. Stansfeld, "Auditory and Non-Auditory Effects of Noise on Health," *Lancet* 383, no. 9925 (2014): 1325–1332; M. Basner, U. Muller, and E. M. Elmenhorst, "Single and Combined Effects of Air, Road, and Rail Traffic Noise on Sleep and Recuperation," *Sleep* 34, no. 1 (2011): 11–23.

9. E. F. Chang and M. M. Merzenich, "Environmental Noise Retards Auditory Cortical Development," *Science* 300, no. 5618 (2003): 498–502; X. Yu, D. H. Sanes, O. Aristizabal, Y. Z. Wadghiri, and D. H. Turnbull, "Large-Scale Reorganization of the Tonotopic Map in Mouse Auditory Midbrain Revealed by MRI," *Proceedings of the National Academy of Sciences of the United States of America* 104, no. 29 (2007): 12193–12198.

10. A. Lahav and E. Skoe, "An Acoustic Gap between the NICU and Womb: A Potential Risk for Compromised Neuroplasticity of the Auditory System in Preterm Infants," *Frontiers in Neuroscience* 8 (2014): 381.

11. E. McMahon, P. Wintermark, and A. Lahav, "Auditory Brain Development in Premature Infants: the Importance of Early Experience," *Annals of the New York Academy of Sciences* 1252 (2012): 17–24.

12. D. E. Anderson and A. D. Patel, "Infants Born Preterm, Stress, and Neurodevelopment in the Neonatal Intensive Care Unit: Might Music Have an Impact?" *Developmental Medicine and Child Neurology* 60, no. 3 (2018): 256–266.

13. A. R. Webb, H. T. Heller, C. B. Benson, and A. Lahav, "Mother's Voice and Heartbeat Sounds Elicit Auditory Plasticity in the Human Brain Before Full Gestation," *Proceedings of the National Academy of Sciences of the United States of America* 112, no. 10 (2015): 3152–3157.

14. S. Arnon, A. Shapsa, L. Forman, R. Regev, S. Bauer, I. Litmanovitz, and T. Dolfin, "Live Music Is Beneficial to Preterm Infants in the Neonatal intensive Care Unit Environment," *Birth* 33, no. 2 (2006): 131–136.

15. X. Zhou, R. Panizzutti, E. de Villers-Sidani, C. Madeira, and M. M. Merzenich, "Natural Restoration of Critical Period Plasticity in the Juvenile and Adult Primary Auditory Cortex," *Journal of Neuroscience* 31, no. 15 (2011): 5625–5634.

16. A. J. Noreña and J. J. Eggermont, "Enriched Acoustic Environment after Noise Trauma Reduces Hearing Loss and Prevents Cortical Map Reorganization," *Journal of Neuroscience* 25, no. 3 (2005): 699–705.

17. M. Pienkowski and J. J. Eggermont, "Long-Term, Partially-Reversible Reorganization of Frequency Tuning in Mature Cat Primary Auditory Cortex Can Be Induced by Passive Exposure to Moderate-Level Sounds," *Hearing Research* 257, nos. 1–2 (2009): 24–40; M. Pienkowski and J. J. Eggermont, "Intermittent Exposure with Moderate-Level Sound Impairs Central Auditory Function of Mature Animals Without Concomitant Hearing Loss," *Hearing Research* 261, no. 1–2 (2010): 30–35; W. Zheng, "Auditory Map Reorganization and Pitch Discrimination in Adult Rats Chronically Exposed to Low-Level Ambient Noise," *Frontiers in Systems Neuroscience* 6 (2012): 65; M. Pienkowski, R. Munguia, and J. J. Eggermont, "Effects of Passive, Moderate-Level Sound Exposure on the Mature Auditory Cortex: Spectral Edges, Spectrotemporal Density, and Real-World Noise," *Hearing Research* 296 (2012): 121–130.

18. E. Hoff, B. Laursen, and K. Bridges, "Measurement and Model Building in Studying the Influence of Socioeconomic Status on Child Development," in *The Cambridge Handbook of Environment in Human Development* (Cambridge: Cambridge University Press, 2012), 590–606.

19. E. Skoe, J. Krizman, and N. Kraus, "The Impoverished Brain: Disparities in Maternal Education Affect the Neural Response to Sound," *Journal of Neuroscience* 33, no. 44 (2013): 17221–17231.

20. B. Hart and T. R. Risley, *Meaningful Differences in the Everyday Experience of Young American Children* (Baltimore: P.H. Brookes, 1995).

21. L. M. Dale, S. Goudreau, S. Perron, M. S. Ragettli, M. Hatzopoulou, and A. Smargiassi, "Socioeconomic Status and Environmental Noise Exposure in Montreal, Canada," *BMC Public Health* 15 (2015): 205.

22. W. H. Mulders, D. Ding, R. Salvi, and D. Robertson, "Relationship between Auditory Thresholds, Central Spontaneous Activity, and Hair Cell Loss after

Acoustic Trauma," *Journal of Comparative Neurology* 519, no. 13 (2011): 2637–47; A. J. Norena and J. J. Eggermont, "Changes in Spontaneous Neural Activity Immediately After an Acoustic Trauma: Implications for Neural Correlates of Tinnitus," *Hearing Research* 183, no. 1–2 (2003): 137–153.

23. J. J. Eggermont, *Tinnitus: Springer Handbook of Auditory Research* (New York: Springer, 2012).

24. M. Attarha, J. Bigelow, and M. M. Merzenich, "Unintended Consequences of White Noise Therapy for Tinnitus—Otolaryngology's Cobra Effect: A Review," *JAMA Otolaryngology—Head and Neck Surgery* 144, no. 10 (2018): 938–943.

25. B. Mazurek, A. J. Szczepek, and S. Hebert, "Stress and Tinnitus," *HNO* 63, no. 4 (2015): 258–265; P. J. Jastreboff and M. M. Jastreboff, "Tinnitus Retraining Therapy (TRT) as a Method for Treatment of Tinnitus and Hyperacusis Patients," *Journal of the American Academy of Audiology* 11 (2000): 162–177.

26. R. Tyler, A. Cacace, C. Stocking, B. Tarver, N. Engineer, J. Martin, A. Deshpande, N. Stecker, M. Pereira, M. Kilgard, C. Burress, D. Pierce, R. Rennaker, and S. Vanneste, "Vagus Nerve Stimulation Paired with Tones for the Treatment of Tinnitus: A Prospective Randomized Double-Blind Controlled Pilot Study in Humans," *Scientific Reports* 7, no. 1 (2017): 11960.

27. W. H. Mulders, D. Ding, R. Salvi, and D. Robertson, "Relationship between Auditory Thresholds, Central Spontaneous Activity, and Hair Cell Loss After Acoustic Trauma," *Journal of Comparative Neurology* 519, no. 13 (2011): 2637–2647; A. J. Norena and J. J. Eggermont, "Changes in Spontaneous Neural Activity Immediately After an Acoustic Trauma: Implications for Neural Correlates of Tinnitus," *Hearing Research* 183, no. 1–2 (2003): 137–153.

28. T. Gioia, *Healing Songs* (Durham, NC: Duke University Press, 2006).

29. G. Hempton and J. Grossmann, *One Square Inch of Silence: One Man's Search for Natural Silence in a Noisy World* (New York: Free Press, 2009).

30. M. A. Denolle and T. Nissen-Meyer, "Quiet Anthropocene, Quiet Earth," *Science* 369, no. 6509 (2020): 1299–1300.

31. G. L. Patricelli and J. L. Blickley, "Avian Communication in Urban Noise: Causes and Consequences of Vocal Adjustment," *Auk* 123, no. 3 (2006): 639–649; J. W. C. Sun and P. A. Narins, "Anthropogenic Sounds Differentially Affect Amphibian Call Rate," *Biological Conservation* 121, no. 3 (2005): 419–27; S. E. Parks, M. Johnson, D. Nowacek, and P. L. Tyack, "Individual Right Whales Call Louder in increased Environmental Noise," *Biology Letters* 7, no. 1 (2011): 33–35.

32. W. E. Wood and S. M. Yezerinac, "Song Sparrow (Melospiza Melodia) Song Varies with Urban Noise," *Auk* 123, no. 3 (2006): 650–659.

33. E. P. Derryberry, J. N. Phillips, G. E. Derryberry, M. J. Blum, and D. Luther, "Singing in a Silent Spring: Birds Respond to a Half-Century Soundscape Reversion during the COVID-19 Shutdown," *Science* 370, no. 6516 (2020): 575–579.

34. A. Fernandez, M. Arbelo, and V. Martin, "No Mass Strandings Since Sonar Ban," *Nature* 497, no. 7449 (2013): 317.

35. M. Waldman, *My Fellow Americans: The Most Important Speeches of America's Presidents, from George Washington to Barack Obama* (Naperville, IL: Sourcebooks, 2010).

36. B. Bosker, "The End of Silence," *Atlantic*, November 2019.

37. A. J. Blood and R. J. Zatorre, "Intensely Pleasurable Responses to Music Correlate with Activity in Brain Regions Implicated in Reward and Emotion," *Proceedings of the National Academy of Sciences USA* 98, no. 20 (2001): 11818–11823; V. N. Salimpoor, I. van Den Bosch, N. Kovacevic, R. R. Mcintosh, A. Dagher, and R. J. Zatorre, "Interactions between the Nucleus Accumbens and Auditory Cortices Predict Music Reward Value," *Science* 340, no. 6129 (2013): 216–219; V. N. Salimpoor, M. Benovoy, K. Larcher, A. Dagher, and R. J. Zatorre, "Anatomically Distinct Dopamine Release During Anticipation and Experience of Peak Emotion to Music," *Nature Neuroscience* 14, no. 2 (2011): 257–262.

38. N. Martinez-Molina, E. Mas-Herrero, A. Rodriguez-Fornells, R. J. Zatorre, and J. Marco-Pallares, "Neural Correlates of Specific Musical Anhedonia," *Proceedings of the National Academy of Sciences of the United States of America* 113, no. 46 (2016): E7337–345.

39. See "Paris Police Step Up Anti-noise Patrols," BBC News, July 25, 2020, https://www.bbc.com/news/av/world-europe-53521561/paris-police-step-up-anti-noise-patrols.

Chapter 12

1. K. J. Cruickshanks, T. L. Wiley, T. S. Tweed, B. E. K. Klein, R. Klein, J. A. Mares-Perlman, and D. M. Nondahl, "Prevalence of Hearing Loss in Older Adults in Beaver Dam, Wisconsin—the Epidemiology of Hearing Loss Study," *American Journal of Epidemiology* 148, no. 9 (1998): 879–86.

2. F. R. Lin, R. Thorpe, S. Gordon-Salant, and L. Ferrucci, "Hearing Loss Prevalence and Risk Factors Among Older Adults in the United States," *Journals of Gerontology Series A: Biological Sciences and Medical Sciences* 66, no. 5 (2011): 582–90.

3. J. F. Willott, "Anatomic and Physiologic Aging: A Behavioral Neuroscience Perspective," *Journal of the American Academy of Audiology* 7, no. 3 (1996): 141–51.

4. S. Anderson and N. Kraus, "The Potential Role of the cABR in Assessment and Management of Hearing Impairment," *International Journal of Otolaryngology* 2013, no. 604729 (2013): 1–10; H. Karawani, K. Jenkins, and S. Anderson, "Restoration of

Sensory Input May Improve Cognitive and Neural Function," *Neuropsychologia* 114 (2018): 203–13; H. Karawani, K. Jenkins, and S. Anderson, "Neural and Behavioral Changes After the Use of Hearing Aids," *Clinical Neurophysiology* 129, no. 6 (2018): 1254–67; K. A. Jenkins, C. Fodor, A. Presacco, and S. Anderson, "Effects of Amplification on Neural Phase Locking, Amplitude, and Latency to a Speech Syllable," *Ear and Hearing* 39, no. 4 (2018): 810–24.

5. J. P. Walton, H. Simon, and R. D. Frisina, "Age-Related Alterations in the Neural Coding of Envelope Periodicities," *Journal of Neurophysiology* 88, no. 2 (2002): 565–78.

6. D. M. Caspary, L. Ling, J. G. Turner, and L. F. Hughes, "Inhibitory Neurotransmission, Plasticity and Aging in the Mammalian Central Auditory System," *Journal of Experimental Biology* 211, no. 11 (2008): 1781–91; D. M. Caspary, L. F. Hughes, and L. L. Ling. "Age-Related GABAA Receptor Changes in Rat Auditory Cortex." *Neurobiology of Aging* 34, no. 5 (2013): 1486–96; J. R. Engle and G. H. Recanzone, "Characterizing Spatial Tuning Functions of Neurons in the Auditory Cortex of Young and Aged Monkeys: A New Perspective on Old Data," *Frontiers in Aging Neuroscience* 4 (2012): 36; D. M. Caspary, T. A. Schatteman, and L. F. Hughes, "Age-Related Changes in the Inhibitory Response Properties of Dorsal Cochlear Nucleus Output Neurons: Role of Inhibitory Inputs," *Journal of Neuroscience* 25, no. 47 (2005): 10952–59; E. de Villers-Sidani, L. Alzghoul, X. Zhou, K. L. Simpson, R. C. Lin, and M. M. Merzenich, "Recovery of Functional and Structural Age-Related Changes in the Rat Primary Auditory Cortex with Operant Training," *Proceedings of the National Academy of Sciences of the USA* 107, no. 31 (2010): 13900–5; B. D. Richardson, L. L. Ling, V. V. Uteshev, and D. M. Caspary, "Reduced GABA(A) Receptor-Mediated Tonic Inhibition in Aged Rat Auditory Thalamus," *Journal of Neuroscience* 33, no. 3 (2013): 1218–27a; D. L. Juarez-Salinas, J. R. Engle, X. O. Navarro, and G. H. Recanzone, "Hierarchical and Serial Processing in the Spatial Auditory Cortical Pathway Is Degraded by Natural Aging," *Journal of Neuroscience* 30, no. 44 (2010): 14795–804.

7. D. M. Caspary, L. Ling, J. G. Turner, and L. F. Hughes, "Inhibitory Neurotransmission, Plasticity and Aging in the Mammalian Central Auditory System," *Journal of Experimental Biology* 211(11): 1781–91; J. H. Grose and S. K. Mamo, "Processing of Temporal Fine Structure as a Function of Age," *Ear and Hearing* 31, no. 6 (2010): 755–60; K. L. Tremblay, M. Piskosz, and P. Souza, "Effects of Age and Age-Related Hearing Loss on the Neural Representation of Speech Cues," *Clinical Neurophysiology* 114, no. 7 (2003): 1332–43; K. C. Harris, M. A. Eckert, J. B. Ahlstrom, and J. R. Dubno, "Age-Related Differences in Gap Detection: Effects of Task Difficulty and Cognitive Ability," *Hearing Research* 264, no. 1–2 (2010): 21–29; J. J. Lister, N. D. Maxfield, G. J. Pitt, and V. B. Gonzalez, "Auditory Evoked Response to Gaps in Noise: Older Adults," *International Journal of Audiology* 50, no. 4 (2011): 211–25; J. P. Walton, "Timing Is Everything: Temporal Processing Deficits in the Aged Auditory Brainstem," *Hearing Research* 264, no. 1–2 (2010): 63–69; L. E. Humes, D. Kewley-Port, D. Fogerty, and

D. Kinney, "Measures of Hearing Threshold and Temporal Processing Across the Adult Lifespan," *Hearing Research* 264, no. 1–2 (2010): 30–40.

8. W. C. Clapp, M. T. Rubens, J. Sabharwal, and A. Gazzaley, "Deficit in Switching between Functional Brain Networks Underlies the Impact of Multitasking on Working Memory in Older Adults," *Proceedings of the National Academy of Sciences of the USA* 108 no. 17 (2011): 7212–17; A. Gazzaley, J. W. Cooney, J. Rissman, and M. D'Esposito, "Top-Down Suppression Deficit Underlies Working Memory Impairment in Normal Aging," *Nature Neuroscience* 8, no. 10 (2005): 1298–300.

9. D. L. Juarez-Salinas, J. R. Engle, X. O. Navarro, and G. H. Recanzone, "Hierarchical and Serial Processing in the Spatial Auditory Cortical Pathway Is Degraded by Natural Aging," *Journal of Neuroscience* 30, no. 44 (2010): 14795–804.

10. R. Peters, "Ageing and the Brain," *Postgraduate Medical Journal* 82, no. 964 (2006): 84–88.

11. T. A. Salthouse, "The Processing-Speed Theory of Adult Age Differences in Cognition," *Psychological Review* 103(3): 403–28; C. T. Albinet, G. Boucard, C. A. Bouquet, and M. Audiffren, "Processing Speed and Executive Functions in Cognitive Aging: How to Disentangle Their Mutual Relationship?" *Brain and Cognition* 79, no. 1 (2012): 1–11; R. Zacks, L. Hasher, and K. Li, "Human Memory," in *Handbook of Aging and Cognition*, ed. F. Craik and T. Salthouse, 293–358 (Mahwah, NJ: Erlbaum, 2000).

12. D. M. Caspary, L. Ling, J. G. Turner, and L. F. Hughes, "Inhibitory Neurotransmission, Plasticity and Aging in the Mammalian Central Auditory System," *Journal of Experimental Biology* 211, no. 11 (2008): 1781–91; D. L. Juarez-Salinas, J. R. Engle, X. O. Navarro, and G. H. Recanzone, "Hierarchical and Serial Processing in the Spatial Auditory Cortical Pathway Is Degraded by Natural Aging," *Journal of Neuroscience* 30, no. 44 (2010): 14795–804; J. J. Lister, R. A. Roberts, and F. L. Lister, "An Adaptive Clinical Test of Temporal Resolution: Age Effects," *International Journal of Audiology* 50, no. 6 (2011): 367–74.

13. T. Salthouse, "Consequences of Age-Related Cognitive Declines," *Annual Review of Psychology* 63 (2012): 201–26.

14. R. Katzman, R. Terry, R. Deteresa, T. Brown, P. Davies, P. Fuld, R. B. Xiong, and A. Peck, "Clinical, Pathological, and Neurochemical Changes in Dementia—a Subgroup with Preserved Mental Status and Numerous Neocortical Plaques," *Annals of Neurology* 23, no. 2 (1988): 138–44.

15. C. M. Tomaino, "Meeting the Complex Needs of Individuals with Dementia Through Music Therapy," *Music and Medicine* 5, no. 4 (2013): 234–41.

16. S. Anderson, A. Parbery-Clark, T. White-Schwoch, and N. Kraus, "Aging Affects Neural Precision of Speech Encoding," *Journal of Neuroscience* 32, no. 41 (2012): 14156–64.

17. B. U. Forstmann, M. Tittgemeyer, E. J. Wagenmakers, J. Derrfuss, D. Impe-
rati, and S. Brown, "The Speed-Accuracy Tradeoff in the Elderly Brain:A Structural
Model-Based Approach," *Journal of Neuroscience* 31, no. 47 (2011): 17242–49; P. H.
Lu, G. J. Lee, E. P. Raven, K. Tingus, T. Khoo, P. M. Thompson, and G. Bartzokis,
"Age-Related Slowing in Cognitive Processing Speed Is Associated with Myelin
Integrity in a Very Healthy Elderly Sample," *Journal of Clinical and Experimental Neu-
ropsychology* 33, no. 10 (2011): 1059–68.

18. S. Anderson, A. Parbery-Clark, T. White-Schwoch, and N. Kraus, "Auditory
Brainstem Response to Complex Sounds Predicts Self-Reported Speech-in-Noise Performance," *Journal of Speech, Language, and Hearing Research* 56, no. 1 (2013): 31–43.

19. H. A. Glick and A. Sharma, "Cortical Neuroplasticity and Cognitive Function in
Early-Stage, Mild-Moderate Hearing Loss: Evidence of Neurocognitive Benefit From
Hearing Aid Use," *Frontiers in Neuroscience* 14 (2020): 93.

20. Max Planck Institute for Human Development and Stanford Center on Longev-
ity, "A Consensus on the Brain Training Industry from the Scientific Community."
http://longevity3.stanford.edu/blog/2014/10/15/the-consensus-on-the-brain
-training-industry-from-the-scientific-community-2/.

21. S. Anderson, T. White-Schwoch, A. Parbery-Clark, and N. Kraus, "Reversal of
Age-Related Neural Timing Delays with Training," *Proceedings of the National Academy of Sciences of the United States of America* 110, no. 11 (2013): 4357–62.

22. S. Anderson, T. White-Schwoch, H. J. Choi, and N. Kraus, "Partial Maintenance
of Auditory-Based Cognitive Training Benefits in Older Adults," *Neuropsychologia* 62
(2014): 286–96.

23. J. Verghese, R. B. Lipton, M. J. Katz, C. B. Hall, C. A. Derby, G. Kuslansky, A. F.
Ambrose, M. Sliwinski, and H. Buschke, "Leisure Activities and the Risk of Dementia
in the Elderly," *New England Journal of Medicine* 348, no. 25 (2003): 2508–16; S. C.
Moore, A. V. Patel, C. E. Matthews, A. Berrington de Gonzalez, Y. Park, H. A. Katki,
M. S. Linet, E. Weiderpass, K. Visvanathan, K. J. Helzlsouer, M. Thun, S. M. Gapstur,
P. Hartge, and I. M. Lee, "Leisure Time Physical Activity of Moderate to Vigorous
intensity and Mortality: A Large Pooled Cohort Analysis," *PLoS Medicine* 9, no. 11
(2012): e1001335.

24. F. R. Lin, E. J. Metter, R. J. O'Brien, S. M. Resnick, A. B. Zonderman, and L. Ferrucci,
"Hearing Loss and Incident Dementia," *Archives of Neurology* 68, no. 2 (2011): 214–20;
R. K. Gurgel, P. D. Ward, S. Schwartz, M. C. Norton, N. L. Foster, and J. T. Tschanz,
"Relationship of Hearing Loss and Dementia: A Prospective, Population-Based Study,"
Otology & Neurotology 35, no. 5 (2014): 775–81; F. R. Lin, K. Yaffe, J. Xia, Q. L. Xue, T.
B. Harris, E. Purchase-Helzner, S. Satterfield, H. N. Ayonayon, L. Ferrucci, E. M. Simon-
sick, and Health ABC Study Group, "Hearing Loss and Cognitive Decline in Older
Adults," *JAMA Internal Medicine* 173, no. 4 (2013): 293–99.

25. R. K. Gurgel, P. D. Ward, S. Schwartz, M. C. Norton, N. L. Foster, and J. T. Tschanz, "Relationship of Hearing Loss and Dementia: A Prospective, Population-Based Study," *Otology & Neurotology* 35, no. 5 (2014): 775–81; C. A. Peters, J. F. Potter, and S. G. Scholer, "Hearing Impairment as a Predictor of Cognitive Decline in Dementia," *Journal of the American Geriatrics Society* 36, no. 11 (1998): 981–86.

26. G. Livingston, A. Sommerlad, V. Orgeta, S. G. Costafreda, J. Huntley, D. Ames, C. Ballard, S. Banerjee, A. Burns, J. Cohen-Mansfield, C. Cooper, N. Fox, L. N. Gitlin, R. Howard, H. C. Kales, E. B. Larson, K. Ritchie, K. Rockwood, E. L. Sampson, Q. Samus, L. S. Schneider, G. Selbaek, L. Teri, and N. Mukadam, "Dementia Prevention, Intervention, and Care," *Lancet* 390, no. 10113 (2017): 2673–2734.

27. G. A. Gates, R. K. Karzon, P. Garcia, J. Peterein, M. Storandt, J. C. Morris, and J. P. Miller, "Auditory Dysfunction in Aging and Senile Dementia of the Alzheimer's Type," *Archives in Neurology* 52, no. 6 (1995): 626–634; G. A. Gates, M. L. Anderson, S. M. McCurry, M. P. Feeney, and E. B. Larson, "Central Auditory Dysfunction as a Harbinger of Alzheimer Dementia," *Archives of Otolaryngology—Head and Neck Surgery* 137, no. 4 (2011): 390–395.

28. B. R. Zendel and C. Alain, "Musicians Experience Less Age-Related Decline in Central Auditory Processing," *Psychology and Aging* 27, no. 2 (2012): 410–17; G. M. Bidelman and C. Alain, "Musical Training Orchestrates Coordinated Neuroplasticity in Auditory Brainstem and Cortex to Counteract Age-Related Declines in Categorical Vowel Perception," *Journal of Neuroscience* 35, no. 3 (2015): 1240–49.

29. B. Pladdy and A. MacKay, "The Relation between Instrumental Musical Activity and Cognitive Aging," *Neuropsychology* 25, no. 3 (2011): 378–86.

30. A. Parbery-Clark, D. L. Strait, S. Anderson, E. Hittner, and N. Kraus, "Musical Experience and the Aging Auditory System: Implications for Cognitive Abilities and Hearing Speech in Noise," *PLoS One* 6, no. 5 (2011): e18082.

31. S. Anderson, A. Parbery-Clark, T. White-Schwoch, and N. Kraus, "Aging Affects Neural Precision of Speech Encoding," *Journal of Neuroscience* 32, no. 41 (2012): 14156–64; A. Parbery-Clark, S. Anderson, E. Hittner, and N. Kraus, "Musical Experience Strengthens the Neural Representation of Sounds Important for Communication in Middle-Aged Adults," *Frontiers in Aging Neuroscience* 4, no. 30 (2012): 1–12.

32. A. Parbery-Clark, D. L. Strait, S. Anderson, E. Hittner, and N. Kraus, "Musical Experience and the Aging Auditory System: Implications for Cognitive Abilities and Hearing Speech in Noise," *PLoS One* 6, no. 5 (2011): e18082; A. Parbery-Clark, S. Anderson, and N. Kraus, "Musicians Change Their Tune: How Hearing Loss Alters the Neural Code," *Hearing Research* 302 (2013): 121–31.

33. E. Skoe and N. Kraus, "A Little Goes a Long Way: How the Adult Brain Is Shaped by Musical Training in Childhood," *Journal of Neuroscience* 32, no. 34 (2012):

11507–10; T. White-Schwoch, K. W. Carr, S. Anderson, D. L. Strait, and N. Kraus, "Older Adults Benefit from Music Training Early in Life: Biological Evidence for Long-Term Training-Driven Plasticity," *Journal of Neuroscience* 33, no. 45 (2012): 17667–74.

34. S. W. Threlkeld, C. A. Hill, G. D. Rosen, and R. H. Fitch, "Early Acoustic Discrimination Experience Ameliorates Auditory Processing Deficits in Male Rats with Cortical Developmental Disruption," *International Journal of Developmental Neuroscience* 27, no. 4 (2009): 321–28; E. C. Sarro and D. H. Sanes, "The Cost and Benefit of Juvenile Training on Adult Perceptual Skill," *Journal of Neuroscience* 31, no. 14 (2011): 5383–91; N. D. Engineer, C. R. Percaccio, P. K. Pandya, R. Moucha, D. L. Rathbun, and M. P. Kilgard, "Environmental Enrichment Improves Response Strength, Threshold, Selectivity, and Latency of Auditory Cortex Neurons," *Journal of Neurophysiology* 92, no. 1 (2004): 73–82.

35. B. Hanna-Pladdy and A. MacKay, "The Relation between Instrumental Musical Activity and Cognitive Aging," *Neuropsychology* 25, no. 3 (2011): 378–86; B. Hanna-Pladdy and B. Gajewski, "Recent and Past Musical Activity Predicts Cognitive Aging Variability: Direct Comparison with General Lifestyle Activities," *Frontiers in Human Neuroscience* 6 (2012): 198.

36. E. de Villers-Sidani, L. Alzghoul, X. Zhou, K. L. Simpson, R. C. Lin, and M. M. Merzenich, "Recovery of Functional and Structural Age-Related Changes in the Rat Primary Auditory Cortex with Operant Training," *Proceedings of the National Academy of Sciences of the USA* 107, no. 31 (2010): 13900–5; E. de Villers-Sidani and M. M. Merzenich, "Lifelong Plasticity in the Rat Auditory Cortex: Basic Mechanisms and Role of Sensory Experience," *Progress in Brain Research* 191 (2011): 119–31; J. M. Cisneros-Franco, L. Ouellet, B. Kamal, and E. de Villers-Sidani, "A Brain Without Brakes: Reduced Inhibition Is Associated with Enhanced but Dysregulated Plasticity in the Aged Rat Auditory Cortex," *eNeuro* 5, no. 4 (2018).

37. E. Dubinsky, E. A. Wood, G. Nespoli, and F. A. Russo, "Short-Term Choir Singing Supports Speech-in-Noise Perception and Neural Pitch Strength in Older Adults with Age-Related Hearing Loss," *Frontiers in Neuroscience* 13 (2019): 1153.

38. B. R. Zendel, G. L. West, S. Belleville, and I. Peretz, "Musical Training Improves the Ability to Understand Speech-in-Noise in Older Adults," *Neurobiology of Aging* 81 (2019): 102–115.

39. J. A. Bugos, "The Effects of Bimanual Coordination in Music Interventions on Executive Functions in Aging Adults," *Frontiers in Integrative Neuroscience* 13 (2019): 68.

40. J. K. Johnson, J. Louhivuori, A. L. Stewart, A. Tolvanen, L. Ross, and P. Era, "Quality of Life (QOL) of Older Adult Community Choral Singers in Finland," *International Psychogeriatrics* 25, no. 7 (2013): 1055–64; J. K. Johnson, A. L. Stewart, M.

Acree, A. M. Napoles, J. D. Flatt, W. B. Max, and S. E. Gregorich, "A Community Choir Intervention to Promote Well-Being Among Diverse Older Adults: Results from the Community of Voices Trial," *Journals of Gerontology Series B: Psychological Sciences and Social Sciences* (2018): https://doi.org/10.1093/geronb/gby132.

41. G. D. Cohen, S. Perlstein, J. Chapline, J. Kelly, K. M. Firth, and S. Simmens, "The Impact of Professionally Conducted Cultural Programs on the Physical Health, Mental Health, and Social Functioning of Older Adults," *Gerontologist* 46, no. 6 (2006): 726–34.

42. J. K. Johnson, J. Louhivuori, A. L. Stewart, A. Tolvanen, L. Ross, and P. Era, "Quality of Life (QOL) of Older Adult Community Choral Singers in Finland," *International Psychogeriatrics* 25, no. 7 (2013): 1055–1064; T. Särkämö, S. Laitinen, A. Numminen, M. Kurki, J. K. Johnson, and P. Rantanen, "Pattern of Emotional Benefits Induced by Regular Singing and Music Listening in Dementia," *Journal of the American Geriatrics Society* 64, no. 2 (2016): 439–440; T. Särkämö, M. Tervaniemi, S. Laitinen, A. Numminen, M. Kurki, J. K. Johnson, and P. Rantanen, "Cognitive, Emotional, and Social Benefits of Regular Musical Activities in Early Dementia: Randomized Controlled Study," *Gerontologist* 54, no. 4 (2014): 634–650.

43. E. Bialystok, F. I. Craik, R. Klein, and M. Viswanathan, "Bilingualism, Aging, and Cognitive Control: Evidence from the Simon Task," *Psychology and Aging* 19, no. 2 (2004): 290–303.

44. T. A. Schweizer, J. Ware, C. E. Fischer, F. I. Craik, and E. Bialystok, "Bilingualism as a Contributor to Cognitive Reserve: Evidence from Brain Atrophy in Alzheimer's Disease," *Cortex* 48, no. 8 (2012): 991–996.

45. E. Woumans, P. Santens, A. Sieben, J. Versijpt, M. Stevens, and W. Duyck, "Bilingualism Delays Clinical Manifestation of Alzheimer's Disease," *Bilingualism: Language and Cognition* 18, no. 3 (2015): 568–574; F. I. Craik, E. Bialystok, and M. Freedman, "Delaying the Onset of Alzheimer Disease: Bilingualism as a Form of Cognitive Reserve," *Neurology* 75, no. 19 (2010): 1726–1729.

Chapter 13

1. H. Kraus and R. P. Hirschland, "Muscular Fitness and Health," *Journal of the American Association for Health, Physical Education, and Recreation* 24, no. 10 (1953): 17–19; H. Kraus and R. P. Hirschland, "Muscular Fitness and Orthopedic Disability," *New York State Journal of Medicine* 54, no. 2 (1954): 212–215.

2. R. H. Boyle, "The Report That Shocked the President," *Sports Illustrated*, August 15, 1955.

3. C. H. Hillman, K. I. Erickson, and A. F. Kramer, "Be Smart, Exercise Your Heart: Exercise Effects on Brain and Cognition," *Nature Reviews Neuroscience* 9, no. 1 (2008): 58–65; M. W. Voss, A. F. Kramer, C. Basak, R. S. Prakash, and B. Roberts, "Are

Expert Athletes 'Expert' in the Cognitive Laboratory? A Meta-Analytic Review of Cognition and Sport Expertise," *Applied Cognitive Psychology* 24, no. 6 (2010): 812–826; F. M. Iaia and J. Bangsbo, "Speed Endurance Training Is a Powerful Stimulus for Physiological Adaptations and Performance Improvements of Athletes," *Scandinavian Journal of Medicine & Science in Sports* 20, Suppl. 2 (2010): 11–23; T. R. Bashore, B. Ally, N. C. van Wouwe, J. S. Neimat, W. P. M. van Den Wildenberg, and S. A. Wylie, "Exposing an 'Intangible' Cognitive Skill Among Collegiate Football Players: II. Enhanced Response Impulse Control," *Frontiers in Psychology* 9 (2018): 1496; Centers for Disease Control and Prevention, *The Association between School Based Physical Activity, Including Physical Education, and Academic Performance* (Atlanta: US Department of Health and Human Services, 2010).

4. B. Draganski, C. Gaser, V. Busch, G. Schuierer, U. Bogdahn, and A. May, "Neuroplasticity: Changes in Grey Matter Induced by Training," *Nature* 427, no. 6972 (2004): 311–312; M. Taubert, B. Draganski, A. Anwander, K. Muller, A. Horstmann, A. Villringer, and P. Ragert, "Dynamic Properties of Human Brain Structure: Learning-Related Changes in Cortical Areas and Associated Fiber Connections," *Journal of Neuroscience* 30, no. 35 (2010): 11670–11667; C. Sampaio-Baptista, J. Scholz, M. Jenkinson, A. G. Thomas, N. Filippini, G. Smit, G. Douaud, and H. Johansen-Berg, "Gray Matter Volume Is Associated with Rate of Subsequent Skill Learning After a Long Term Training Intervention," *Neuroimage* 96 (2014): 158–166; T. R. Bashore, B. Ally, N. C. van Wouwe, J. S. Neimat, W. P. M. van Den Wildenberg, and S. A. Wylie, "Exposing an 'Intangible' Cognitive Skill Among Collegiate Football Players: II. Enhanced Response Impulse Control," *Frontiers in Psychology* 9 (2018): 1496.

5. I. A. McKenzie, D. Ohayon, H. Li, J. P. de Faria, B. Emery, K. Tohyama, and W. D. Richardson, "Motor Skill Learning Requires Active Central Myelination," *Science* 346, no. 6207 (2014): 318–322.

6. T. Takeuchi, "Auditory Information in Playing Tennis," *Perceptual and Motor Skills* 76, no. 3, pt. 2 (1993): 1323–1328; C. Kennel, L. Streese, A. Pizzera, C. Justen, T. Hohmann, and M. Raab, "Auditory Reafferences: The Influence of Real-Time Feedback on Movement Control," *Frontiers in Psychology* 6 (2015): 69; F. Sors, M. Murgia, I. Santoro, V. Prpic, A. Galmonte, and T. Agostini, "The Contribution of Early Auditory and Visual Information to the Discrimination of Shot Power in Ball Sports," *Psychology of Sport and Exercise* 31 (2017): 44–51; M. Murgia, T. Hohmann, A. Galmonte, M. Raab, and T. Agostini, "Recognising One's Own Motor Actions Through Sound: The Role of Temporal Factors," *Perception* 41, no. 8 (2012): 976–987; I. Camponogara, M. Rodger, C. Craig, and P. Cesari, "Expert Players Accurately Detect an Opponent's Movement Intentions Through Sound Alone," *Journal of Experimental Psychology: Human Perception and Performance* 43, no. 2 (2017): 348–359; N. Schaffert, T. B. Janzen, K. Mattes, and M. H. Thaut, "A Review on the Relationship between Sound and Movement in Sports and Rehabilitation," *Frontiers in Psychology* 10 (2019): 244.

7. J. Krizman, T. Lindley, S. Bonacina, D. Colegrove, T. White-Schwoch, and N. Kraus, "Play Sports for a Quieter Brain: Evidence from Division I Collegiate Athletes," *Sports Health* 12, no. 2 (2020): 154–158.

8. E. Skoe, J. Krizman, and N. Kraus, "The Impoverished Brain: Disparities in Maternal Education Affect the Neural Response to Sound," *Journal of Neuroscience* 33, no. 44 (2013): 17221–17231; H. Luo, E. Pace, X. Zhang, and J. Zhang, "Blast-Induced Tinnitus and Spontaneous Activity Changes in the Rat Inferior Colliculus," *Neuroscience Letters* 580 (2014): 47–51; W. H. Mulders and D. Robertson, "Development of Hyperactivity After Acoustic Trauma in the Guinea Pig Inferior Colliculus," *Hearing Research* 298 (2013): 104–108.

9. C. H. Hillman, K. I. Erickson, and A. F. Kramer, "Be Smart, Exercise Your Heart: Exercise Effects on Brain and Cognition," *Nature Reviews Neuroscience* 9, no. 1 (2008): 58–65; S. E. Fox, P. Levitt, and C. A. Nelson, "How the Timing and Quality of Early Experiences Influence the Development of Brain Architecture," *Child Development* 81, no. 1 (2010): 28–40.

10. Centers for Disease Control and Prevention, "Nonfatal Traumatic Brain injuries Related to Sports and Recreation Activities Among Persons Aged <=19 Years—United States, 2001–2009," *Morbidity and Mortality Weekly Report* 60, no. 39 (2011): 1337–1342.

11. N. Kounang, "Former NFLers Call for End to Tackle Football for Kids," *CNN Health*, March 18, 2018, https://www.cnn.com/2018/01/18/health/nfl-no-tackle-football-kids/index.html.

12. L. S. M, Johnson, "Return to Play Guidelines Cannot Solve the Football-Related Concussion Problem," *Journal of School Health* 82, no. 4 (2012): 180–185.

13. H. S. Martland, "Punch Drunk," *Journal of the American Medical Association* 91 (1928): 1103–1107.

14. A. P. Kontos, T. Covassin, R. J. Elbin, and T. Parker, "Depression and Neurocognitive Performance After Concussion Among Male and Female High School and Collegiate Athletes," *Archives of Physical Medicine and Rehabilitation* 93, no. 10 (2012): 1751–1756; R. D. Moore, W. Sauve, and D. Ellemberg, "Neurophysiological Correlates of Persistent Psycho-Affective Alterations in Athletes with a History of Concussion," *Brain Imaging and Behavior* 10 (2016): 1108; L. M. Mainwaring, M. Hutchison, S. M. Bisschop, P. Comper, and D. W. Richards, "Emotional Response to Sport Concussion Compared to ACL Injury," *Brain Injury* 24, no. 4 (2010): 589–597.

15. B. M. Asken, M. J. Sullan, S. T. DeKosky, M. S. Jaffee, and R. M. Bauer, "Research Gaps and Controversies in Chronic Traumatic Encephalopathy: A Review," *JAMA Neurology* 74, no. 10 (2017): 1255–1262.

16. J. Mez, D. H. Daneshvar, P. T. Kiernan, B. Abdolmohammadi, V. E. Alvarez, B. R. Huber, M. L. Alosco, et al., "Clinicopathological Evaluation of Chronic Traumatic

Encephalopathy in Players of American Football," *Journal of the American Medical Association* 318, no. 4 (2017): 360–370.

17. L. de Beaumont, D. Mongeon, S. Tremblay, J. Messier, F. Prince, S. Leclerc, M. Lassonde, and H. Theoret, "Persistent Motor System Abnormalities in Formerly Concussed Athletes," *Journal of Athletic Training* 46, no. 3 (2017): 234–240; D. M. Bernstein, "Information Processing Difficulty Long After Self-Reported Concussion," *Journal of the International Neuropsychological Society* 8, no. 5 (2002): 673–682; R. D. Moore, S. P. Broglio, and C. H. Hillman, "Sport-Related Concussion and Sensory Function in Young Adults," *Journal of Athletic Training* 49, no. 1 (2014): 36–41; M. B. Pontifex, P. M. O'Connor, S. P. Broglio, and C. H. Hillman, "The Association between Mild Traumatic Brain Injury History and Cognitive Control," *Neuropsychologia* 47, no. 14 (2009): 3210–3216; R. D. Moore, C. H. Hillman, and S. P. Broglio, "The Persistent Influence of Concussive Injuries on Cognitive Control and Neuroelectric Function," *Journal of Athletic Training* 49, no. 1 (2014): 24–35; H. G. Belanger and R. D. Vanderploeg, "The Neuropsychological Impact of Sports-Related Concussion: A Meta-Analysis," *Journal of the International Neuropsychological Society* 11, no. 4 (2005): 345–357; R. S. Moser, P. Schatz, and B. D. Jordan, "Prolonged Effects of Concussion in High School Athletes," *Neurosurgery* 57, no. 2 (2005): 300–306; G. L. Iverson, M. Gaetz, M. R. Lovell, and M. W. Collins, "Cumulative Effects of Concussion in Amateur Athletes," *Brain Injury* 18, no. 5 (2004): 433–443.

18. Arnold Starr, https://www.arnoldstarrart.com.

19. C. Grillon, R. Ameli, and W. M. Glazer, "Brainstem Auditory-Evoked Potentials to Different Rates and Intensities of Stimulation in Schizophrenics," *Biological Psychiatry* 28, no. 9 (1990): 819–823; J. Källstrand, S. F. Nehlstedt, M. L. Sköld, and S. Nielzén, "Lateral Asymmetry and Reduced Forward Masking Effect in Early Brainstem Auditory Evoked Responses in Schizophrenia," *Psychiatry Research* 196, no. 2–3 (2012): 188–193; E. Lahat, E. Avital, J. Barr, M. Berkovitch, A. Arlazoroff, and M. Aladjemm, "BAEP Studies in Children with Attention Deficit Disorder," *Developmental Medicine and Child Neurology* 37, no. 2 (1995): 119–123; S. Otto-Meyer, J. Krizman, T. White-Schwoch, and N. Kraus, "Children with Autism Spectrum Disorder Have Unstable Neural Responses to Sound," *Experimental Brain Research* 236, no. 3 (2018): 733–743; N. M. Russo, E. Skoe, B. Trommer, T. Nicol, S. Zecker, A. Bradlow, and N. Kraus, "Deficient Brainstem Encoding of Pitch in Children with Autism Spectrum Disorders," *Clinical Neurophysiology* 119, no. 8 (2008): 1720–1731; N. M. Russo, T. G. Nicol, B. L. Trommer, S. G. Zecker, and N. Kraus, "Brainstem Transcription of Speech Is Disrupted in Children with Autism Spectrum Disorders," *Developmental Science* 12, no. 4 (2009): 557–567; G. M. Bidelman, J. E. Lowther, S. H. Tak, and C. Alain, "Mild Cognitive Impairment Is Characterized by Deficient Brainstem and Cortical Representations of Speech," *Journal of Neuroscience* 37, no. 13 (2017): 3610–3620; H. Tachibana, M. Takeda, and M. Sugita, "Brainstem Auditory Evoked Potentials in Patients with Multi-Infarct Dementia and Dementia of the Alzheimer Type,"

International Journal of Neuroscience 48, no. 3–4 (1989): 325–331; H. Nakamura, S. Takada, R. Shimabuku, M. Matsuo, T. Matsuo, and H. Negishi, "Auditory Nerve and Brainstem Responses in Newborn Infants with Hyperbilirubinemia," *Pediatrics* 75, no. 4 (1985): 703–8; V. Wahlström, F. Åhlander, and R. Wynn, "Auditory Brainstem Response as a Diagnostic Tool for Patients Suffering from Schizophrenia, Attention Deficit Hyperactivity Disorder, and Bipolar Disorder: Protocol," *JMIR Research Protocols* 4, no. 1 (2015): e16; H. Tachibana, M. Takeda, and M. Sugita, "Short-Latency Somatosensory and Brainstem Auditory Evoked Potentials in Patients with Parkinson's Disease," *International Journal of Neuroscience* 44, no. 3–4 (1989): 321–326; G. Paludetti, F. Ottaviani, V. Gallai, A. Tassoni, and M. Maurizi, "Auditory Brainstem Responses (ABR) in Multiple Sclerosis." *Scandinavian Audiology* 14, no. 1 (1985): 27–34; T. White-Schwoch, A. K. Magohe, A. M. Fellows, C. C. Rieke, B. Vilarello, T. Nicol, E. R. Massawe, N. Moshi, N. Kraus, and J. C. Buckey, "Auditory Neurophysiology Reveals Central Nervous System Dysfunction in HIV-Infected Individuals," *Clinical Neurophysiology* 131 (2020): 1827–1832; E. Castello, N. Baroni, and E. Pallestrini, "Neurotological Auditory Brain Stem Response Findings in Human Immunodeficiency Virus-Positive Patients without Neurologic Manifestations," *Annals of Otology, Rhinology, and Laryngology* 107, no. 12 (1988): 1054–1060.

20. P. McCrory, W. Meeuwisse, J. Dvorak, M. Aubry, J. Bailes, S. Broglio, R. C. Cantu, et al., "Consensus Statement on Concussion in Sport—the 5th international Conference on Concussion in Sport Held in Berlin, October 2016," *British Journal of Sports Medicine* 51 (2017): 838–847.

21. F. J. Gallun, A. C. Diedesch, L. R. Kubli, T. C. Walden, R. L. Folmer, M. S. Lewis, D. J. McDermott, S. A. Fausti, and M. R. Leek, "Performance on Tests of Central Auditory Processing by individuals Exposed to High-Intensity Blasts," *Journal of Rehabilitation Research and Development* 49, no. 7 (2012): 1005–1025.

22. E. C. Thompson, J. Krizman, T. White-Schwoch, T. Nicol, C. R. LaBella, and N. Kraus, "Difficulty Hearing in Noise: A Sequela of Concussion in Children," *Brain Injury* 32, no. 6 (2018): 763–769; C. Turgeon, F. Champoux, F. Lepore, S. Leclerc, and D. Ellemberg, "Auditory Processing After Sport-Related Concussions," *Ear and Hearing* 32, no. 5 (2011): 667–70; P. O. Bergemalm and B. Lyxell, "Appearances Are Deceptive? Long-Term Cognitive and Central Auditory Sequelae from Closed Head Injury," *International Journal of Audiology* 44, no. 1 (2005): 39–49; J. L. Cockrell and S. A. Gregory, "Audiological Deficits in Brain-Injured Children and Adolescents," *Brain Injury* 6, no. 3 (1992): 261–266.

23. L. A. Nelson, M. Macdonald, C. Stall, and R. Pazdan, "Effects of Interactive Metronome Therapy on Cognitive Functioning after Blast-Related Brain Injury: A Randomized Controlled Pilot Trial," *Neuropsychology* 27, no. 6 (2013): 666–679.

24. C. C. Giza and D. A. Hovda, "The New Neurometabolic Cascade of Concussion," *Neurosurgery* 75, suppl. 4 (2014): S24–33.

25. Y. Aoki, R. inokuchi, M. Gunshin, N. Yahagi, and H. Suwa, "Diffusion Tensor Imaging Studies of Mild Traumatic Brain Injury: A Meta-Analysis," *Journal of Neurology, Neurosurgery, and Psychiatry* 83, no. 9 (2012): 870–876.

26. A. A. Hirad, J. J. Bazarian, K. Merchant-Borna, F. E. Garcea, S. Heilbronner, D. Paul, E. B. Hintz, et al., "A Common Neural Signature of Brain Injury in Concussion and Subconcussion," *Science Advances* 5, no. 8 (2019): eaau3460.

27. M. Thériault, L. De Beaumont, N. Gosselin, M. Filipinni, and M. Lassonde, "Electrophysiological Abnormalities in Well Functioning Multiple Concussed Athletes," *Brain Injury* 23, no. 11 (2009): 899–906; S. J. Segalowitz, D. M. Bernstein, and S. Lawson, "P300 Event-Related Potential Decrements in Well-Functioning University Students with Mild Head Injury," *Brain and Cognition* 45, no. 3 (2001): 342–356; R. Pratap-Chand, M. Sinniah, and F. A. Salem, "Cognitive Evoked Potential (P300): A Metric for Cerebral Concussion," *Acta Neurologica Scandinavica* 78, no. 3 (1988): 185–189; N. Gosselin, M. Thériault, S. Leclerc, J. Montplaisir, and M. Lassonde, "Neurophysiological Anomalies in Symptomatic and Asymptomatic Concussed Athletes," *Neurosurgery* 58, no. 6 (2006): 1151–1161.

28. R. M. Amanipour, R. D. Frisina, S. A. Cresoe, T. J. Parsons, Z. Xiaoxia, C. V. Borlongan, and J. P. Walton, "Impact of Mild Traumatic Brain Injury on Auditory Brain Stem Dysfunction in Mouse Model," *Conference Proceedings: Annual International Conference of the IEEE Engineering in Medicine and Biology Society*, (2016): 1854–1857; J. H. Noseworthy, J. Miller, T. J. Murray, and D. Regan, "Auditory Brainstem Responses in Postconcussion Syndrome," *Archives of Neurology* 38, no. 5 (1981): 275–278; F. Ottaviani, G. Almadori, A. B. Calderazzo, A. Frenguelli, and G. Paludetti, "Auditory Brain-Stem (ABRs) and Middle Latency Auditory Responses (MLRs) in the Prognosis of Severely Head-Injured Patients," *Electroencephalography and Clinical Neurophysiology* 65, no. 3 (1986): 196–202; A. Matsumura, I. Mitsui, S. Ayuzawa, S. Takeuchi, and T. Nose, "Prediction of the Reversibility of the Brain Stem Dysfunction in Head injury Patients: MRI and Auditory Brain Stem Response Study," in *Recent Advances in Neurotraumatology*, ed. N. Nakamura, T. Hashimoto, and M. Yasue (Tokyo: Springer Japan, 1993), 192–195.

29. S. K. Munjal, N. K. Panda, and A. Pathak, "Relationship between Severity of Traumatic Brain Injury (TBI) and Extent of Auditory Dysfunction," *Brain Injury* 24, no. 3 (2010): 525–532; Y. Haglund and H. E. Persson, "Does Swedish Amateur Boxing Lead to Chronic Brain Damage? 3. A Retrospective Clinical Neurophysiological Study," *Acta Neurologica Scandinavica* 82, no. 6 (1990): 353–360; C. Nölle, I. Todt, R. O. Seidl, and A. Ernst, "Pathophysiological Changes of the Central Auditory Pathway After Blunt Trauma of the Head," *Journal of Neurotrauma* 21, no. 3 (2004): 251–258.

30. E. C. Thompson, J. Krizman, T. White-Schwoch, T. Nicol, C. R. LaBella, and N. Kraus, "Difficulty Hearing in Noise: a Sequela of Concussion in Children," *Brain Injury* 32, no. 6 (2018): 763–769.

31. N. Kraus, E. C. Thompson, J. Krizman, K. Cook, T. White-Schwoch, and C. R. LaBella, "Auditory Biological Marker of Concussion in Children," *Scientific Reports* 6 (2016): 39009.

32. G. Rauterkus, D. Moncrieff, G. Stewart, and E. Skoe, "Baseline, Retest, and Post-injury Profiles of Auditory Neural Function in Collegiate Football Players," *International Journal of Audiology* (2021) https://doi.org/10.1080/14992027.2020. 1860261; K. R. Vander Werff and B. Rieger, "Brainstem Evoked Potential Indices of Subcortical Auditory Processing After Mild Traumatic Brain Injury," *Ear and Hearing* 38, no. 4 (2017): e200–214.

33. J. P. L. Brokx and S. G. Nooteboom, "Intonation and the Perceptual Separation of Simultaneous Voices," *Journal of Phonetics* 10, no. 1 (1982): 23–36; V. Summers and M. R. Leek, "F0 Processing and the Separation of Competing Speech Signals by Listeners with Normal Hearing and with Hearing Loss," *Journal of Speech, Language, and Hearing Research* 41, no. 6 (1998): 1294–1306.

34. N. Kraus, T. Lindley, D. Colegrove, J. Krizman, S. Otto-Meyer, E. C. Thompson, and T. White-Schwoch, "The Neural Legacy of a Single Concussion," *Neuroscience Letters* 646 (2017): 21–23.

35. S. Abrahams, S. M. Fie, J. Patricios, M. Posthumus, and A. V. September, "Risk Factors for Sports Concussion: An Evidence-Based Systematic Review," *British Journal of Sports Medicine* 48, no. 2 (2014): 91–97.

36. T. White-Schwoch, J. Krizman, K. McCracken, J. K. Burgess, E. C. Thompson, T. Nicol, N. Kraus, and C. R. LaBella, "Baseline Profiles of Auditory, Vestibular, and Visual Functions in Youth Tackle Football Players," *Concussion* 4, no. 4 (2020): CNC66; T. White-Schwoch, J. Krizman, K. McCracken, J. K. Burgess, E. C. Thompson, T. Nicol, C. R. LaBella, and N. Kraus, "Performance on Auditory, Vestibular, and Visual Tests Is Stable Across Two Seasons of Youth Tackle Football," *Brain Injury* 34 (2020): 236–244.

Chapter 14

1. P. Weinberger and C. Burton, "The Effect of Sonication on the Growth of Some Tree Seeds," *Canadian Journal of Forest Research–Revue Canadienne De Recherche Forestiere* 11, no. 4 (1981): 840–844.

2. H. Takahashi, H. Suge, and T. Kato, "Growth Promotion by Vibration At 50 Hz in Rice and Cucumber Seedlings," *Plant and Cell Physiology* 32, no. 5 (1991): 729–732.

3. M. Gagliano, M. Grimonprez, M. Depczynski, and M. Renton, "Tuned In: Plant Roots Use Sound to Locate Water," *Oecologia* 184, no. 1 (2017): 151–160.

4. M. Gagliano, S. Mancuso, and D. Robert, "Towards Understanding Plant Bioacoustics," *Trends in Plant Science* 17, no. 6 (2012): 323–325.

5. S. Buchmann, "Pollination in the Sonoran Desert Region," in *A Natural History of the Sonoran Desert*, ed. M. A. Dimmit, P. W. Comus, S. J. Phillips, and L. M. Brewer (Oakland: University of California Press, 2015), 124–129.

6. T. A. C. Gordon, A. N. Radford, I. K. Davidson, K. Barnes, K. McCloskey, S. L. Nedelec, M. G. Meekan, M. I. McCormick, and S. D. Simpson, "Acoustic Enrichment Can Enhance Fish Community Development on Degraded Coral Reef Habitat," *Nature Communications* 10, no. 1 (2019): 5414.

7. A. T. Woods, E. Poliakoff, D. M. Lloyd, J. Kuenzel, R. Hodson, H. Gonda, J. Batchelor, G. B. Dijksterhuis, and A. Thomas, "Effect of Background Noise on Food Perception," *Food Quality and Preference* 22, no. 1 (2011): 42–47.

8. C. Spence, C. Michel, and B. Smith, "Airline Noise and the Taste of Umami," *Flavour* 3, no. 2 (2014): 1–4.

9. M. Cobb, *The Idea of the Brain: The Past and Future of Neuroscience* (New York: Basic Books, 2020); V. S. Ramachandran and S. Blakeslee, *Phantoms in the Brain: Human Nature and the Architecture of the Mind* (New York: William Morrow, 1998).

10. A. D. Patel, "Evolutionary Music Cognition: Cross-Species Studies," in *Foundations in Music Psychology: Theory and Research*, ed. P. J. Rentfrow and D. Levitin (Cambridge, MA: MIT Press, 2019): 459–501.

11. J. Blacking, *How Musical Is Man?* (Seattle: University of Washington Press, 1973).

12. G. Gigerenzer, *Gut Feelings: The Intelligence of the Unconscious* (New York: Viking, 2007).

13. R. G. Geen, "Effects of Attack and Uncontrollable Noise on Aggression," *Journal of Research in Personality* 12, no. 1 (1978): 15–29.

Index

Page numbers in italic type indicate figures.